Angiogenesis Modulations in Health and Disease

Shaker A. Mousa • Paul J. Davis

Editors

Angiogenesis Modulations in Health and Disease

Practical Applications
of Pro- and Anti-angiogenesis Targets

 Springer

Editors
Shaker A. Mousa
The Pharmaceutical Research Institute at
   Albany College of Pharmacy and Health
   Sciences
Rensselaer, NY, USA

Paul J. Davis
The Pharmaceutical Research Institute at
   Albany College of Pharmacy and Health
   Sciences
Rensselaer, NY, USA

ISBN 978-94-007-6466-8   ISBN 978-94-007-6467-5 (eBook)
DOI 10.1007/978-94-007-6467-5
Springer Dordrecht Heidelberg New York London

Library of Congress Control Number: 2013936733

Printed on acid-free paper

Springer is part of Springer Science+Business Media (www.springer.com)

Contents

Part II   Anti-angiogenesis Targets and Clinical Applications

Contributors

Laila H. Anwar Albany College of Pharmacy and Health Sciences, Albany, NY, USA

Hugo R. Arias Department of Medical Education, College of Medicine, California North State University, Elk Grove, CA, USA

Sujit Basu Department of Pathology and Arthur G. James Comprehensive Cancer Center, Ohio State University, Columbus, OH, USA

Dhruba J. Bharali The Pharmaceutical Research Institute at Albany College of Pharmacy and Health Sciences, Rensselaer, NY, USA

Robert C. Block Department of Public Health Sciences, Division of Cardiology, Department of Medicine, University of Rochester School of Medicine and Dentistry, Rochester, NY, USA

Partha Sarathi Dasgupta Signal Transduction and Biogenic Amines Department, Chittaranjan National Cancer Institute, Kolkata, India

Faith B. Davis The Pharmaceutical Research Institute at Albany College of Pharmacy and Health Sciences, Rensselaer, NY, USA

Paul J. Davis The Pharmaceutical Research Institute at Albany College of Pharmacy and Health Sciences, Rensselaer, NY, USA

Department of Medicine, Albany Medical Center, Albany, NY, USA

Dan G. Duda Steele Laboratory for Tumor Biology, Department of Radiation Oncology, Massachusetts General Hospital, Harvard Medical School, Boston, MA, USA

Steve Georas Division of Pulmonary and Critical Care Medicine, Department of Medicine, University of Rochester School of Medicine and Dentistry, Rochester, NY, USA

Hung-Yun Lin Institute of Cancer Biology and Drug Discovery, Taipei Medical University, Taipei, Taiwan

Mary K. Luidens Department of Medicine, Albany Medical College, Albany, NY, USA

Shaker A. Mousa The Pharmaceutical Research Institute at Albany College of Pharmacy and Health Sciences, Rensselaer, NY, USA

Mathangi Srinivasan The Pharmaceutical Research Institute at Albany College of Pharmacy and Health Sciences, Rensselaer, NY, USA

Murat Yalcin Veterinary Medicine Faculty, Department of Physiology, Uludag University, Gorukle, Bursa, Turkey

The Pharmaceutical Research Institute at Albany College of Pharmacy and Health Sciences, Rensselaer, NY, USA

Introduction

Our intent in this textbook is to facilitate development of agents with clinical potential for modulating new blood vessel formation. We provide overviews of strategies in pro- and anti-angiogenesis, discuss certain classical and a number of recently emphasized targets in angiogenesis—such as several integrins—and we review the actions of classes of agents only recently appreciated to stimulate or inhibit angiogenesis. Discussed here is the impact of novel nanotechnology-based formulations of blood vessel-targeted drugs on the actions of these agents. We also review the assets and certain limitations of assays for angiogenesis and cover the challenging subject of biomarkers of new blood vessel formation. Finally, we speculate about new directions in which modulation of angiogenesis may proceed. We did not intend to provide an extensive overview of history and discovery in physiological and pathologic angiogenesis, but rather we present the current strategies in the modulation of angiogenesis in health and diseases.

The feasibility of clinically important modulation of angiogenesis is a legacy of Dr. Judah Folkman's [1]. The concept moved through the three requisite phases of a newly appreciated truth that Schopenhauer described, namely, dismissal, then strenuous opposition and, finally, the declaration that the new truth was self-evident from the start. The motivation to identify steps in angiogenesis that could be manipulated was initially driven in Folkman and others by the desire to interrupt the vascularization of cancers. Potential anti-angiogenesis agents have emerged in concert with vascular growth factor discovery and the uncovering of steps in the molecular bases of growth factor action. A model case is that of vascular endothelial growth factor (VEGF), where antibodies or a trap have targeted VEGF, itself, and other agents have been directed at the VEGF receptor or at the kinases associated with the receptor. Vascular growth factors in addition to VEGF are of course now known to exist. Thus, it is not surprising that clinical effectiveness is variable from one type of cancer to another of a single anti-angiogenesis agent that targets a single apparent point of vulnerability in the mechanism of action of a single vascular growth factor.

The concept of combining several anti-angiogenesis agents that target individual vascular growth factors—VEGF, basic fibroblast growth factor (bFGF) and epidermal growth factor (EGF)—in one therapeutic modality is attractive and, so far, impractical.

That is, the individual agents are expensive to produce and the intellectual property is the province of different companies. An interesting strategy we discuss in several chapters in this text is looking at several endogenous human hormones that only recently have been seen to have angiogenesis-relevant actions that affect more than one vascular growth factor. The prototypical hormone in this context is an anti-angiogenic iodothyronine analogue of thyroid hormone that acts via the cell surface receptor integrin $\alpha v\beta 3$.

Historically, the initial wave of therapeutic interest in anti-angiogenesis has been understandably followed by the search for pro-angiogenesis agents to be administered in settings of ischemia, preferably via local applications to avoid systemic spillover that might lead to increased risk of excessive vascular activation in patients at risk of cancer. The targets of course may be the same as those of anti-angiogenesis strategies where such targets in blood vessel cells have bidirectional capabilities. Several chapters in this textbook explore interesting and more recently appreciated pro-angiogenesis targets, including those subject to modulation by non-neuronal nicotinic acetylcholine receptors and certain receptors subject to modulation by arachidonic acid-derived lipids. The various pro-angiogenesis and anti-angiogenesis strategies described in this book and beyond might benefit greatly from novel nano-formulation approaches.

The co-editors are very grateful to a substantial group of collaborators who contributed as co-authors and discussants to the realization of the textbook. The editorial contributions of Dr. Kelly Keating to each chapter were thoughtful and essential. We also appreciate the assistance of Ilse Hensen and Ganesan Divya at Springer in the completion of this project.

Shaker A. Mousa, PhD, MBA, FACC, FACB
Paul J. Davis, MD

Reference

1. Folkman J (2007) Angiogenesis: an organizing principle for drug discovery? Nat Rev Drug Discov 6(4):273–286

Chapter 1
Angiogenesis Assays: An Appraisal of Current Techniques

Shaker A. Mousa and Paul J. Davis

Abstract A number of satisfactory methods exist for experimentally estimating the pro- or anti-angiogenic activity of growth factors and pharmaceuticals. We review here a selected group of widely used *in vivo* and *in vitro* assays for angiogenesis. The *in vivo* assays of angiogenesis include the chick chorioallantoic membrane, the zebrafish, tumor xenografts in the mouse and the Matrigel® plug. *In vitro* assays included here are aortic ring, sprouting and tube formation assays. No method can entirely satisfactorily reproduce human angiogenesis, and thus a risk exists of securely defining pro- or anti-angiogenesis in these models and of subsequent failure to establish clinical effectiveness. On the other hand, most of these assays are readily reproducible, respond to standard pro-angiogenesis factors such as growth factors and angiogenesis inhibitors, and have acceptable cost.

In vivo and *in vitro* assays of angiogenesis are essential to development of pharmaceuticals that are pro- or anti-angiogenic and to studies of naturally occurring factors that modulate neovascularization (Table 1.1). The ideal assay is reproducible, quantitative, inexpensive, rapid and highly relevant to angiogenesis in human subjects. 'Relevance' means that the process of new vessel formation in the model responds to addition of human vascular growth factors and involves structural proteins, e.g., integrins, of the endothelial cell plasma membrane that are comparable to those found in human cells. Assays that can accommodate tumor cells and permit such cells to modulate blood

S.A. Mousa (✉)
The Pharmaceutical Research Institute at Albany College of Pharmacy and Health Sciences, Rensselaer, NY, USA
e-mail: shaker.mousa@acphs.edu

P.J. Davis
The Pharmaceutical Research Institute at Albany College of Pharmacy and Health Sciences, Rensselaer, NY, USA

Department of Medicine, Albany Medical Center, Albany, NY, USA
e-mail: pdavis.ordwayst@gmail.com

S.A. Mousa and P.J. Davis (eds.), *Angiogenesis Modulations in Health and Disease: Practical Applications of Pro- and Anti-angiogenesis Targets*, DOI 10.1007/978-94-007-6467-5_1, © Springer Science+Business Media Dordrecht 2013

Table 1.1 Angiogenesis models

I. *In vitro* models	II. *In vivo* models
Cultured EC on different substratum:	Matrigel in mice
Matrigel	Chick chorioallantoic membrane (CAM) assay
Collagen/fibronectin	Rabbit cornea
Laminin	Hypoxia/ischemia-induced retinal/iris NV in rats,
Fibrin or gelatin	mice, primate
	Laser-induced choroidal NV
Sprouting from aortic rings	Human skin/human tumor transplanted on SCID mice
	Tumor metastatic models in mice

vessel formation that is intrinsic to the assay are attractive to investigators who study cancer-associated angiogenesis. For example, the chick chorioallantoic membrane (CAM) is immunotolerant of tumor cells from other species during the early phase of embryogenesis upon which the CAM assay of angiogenesis in avian vessels is based.

An issue to be considered at the outset of assays of blood vessel formation is whether the studies are to be of (developmental) vasculogenesis, or of new blood vessel development that is unrelated to development and involves differentiation of endothelial cells from angioblasts prior to vessel sprouting (angiogenesis). As will be noted in the *in vivo* methods discussed below, the CAM assay examines *angiogenesis* away from the developing chick embryo that is present in the model, whereas the zebrafish embryo assay emphasizes *vasculogenesis* but includes readily identifiable vessels that represent angiogenesis.

A number of reviews of angiogenesis assays [1–5] and their history [6] are available. We discuss below the features of a selected group of assays that are widely used and that we believe can be reproduced in most laboratories. All bear the risk of failure to accurately predict the clinical response of angiogenesis-relevant pharmaceuticals because of the phylogenetic leap from animal models to human subjects.

In Vivo Assays

Chick Chorioallantoic Membrane

The CAM assay was introduced by Folkman and co-workers in 1974 [7] and is the most commonly used *in vivo* measurement of angiogenesis [8–10]. A principal method in use is the cutting of a re-sealable window in the shell of the egg containing a developing embryo, insertion of test substance(s) on filter discs and/or xenograft through the window and monitoring the progress of new blood vessel formation for 3 days or longer. The chorioallantoic membrane is oxygen tension-sensitive and thus the shell window is sealed between manipulations. Angiogenesis is scored—number of vessels, branch points—by examining microscope visualization of the membrane about the site of the test substance filter disc or xenograft, or both, and by application of a software template to images of the membrane (see Fig. 1.1; [11]). The latter may

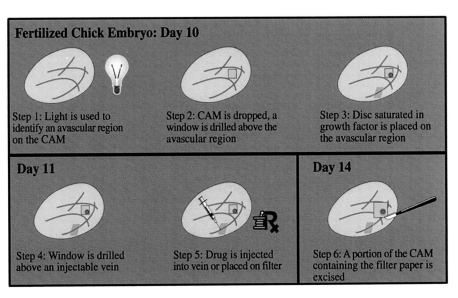

Fig. 1.1 Stepwise protocol for growth factor–induced angiogenesis in the chick chorioallantoic membrane (CAM) model

be automated. A variant of the CAM involves explanting the embryo and CAM without the shell to a Petri dish and carrying out the quantitation of angiogenesis.

The CAM method is reproducible, inexpensive and well-suited to large-scale screening of test drugs, antibodies or growth factors. Membrane samples may be harvested for biochemical analysis, e.g., signal transducing enzyme activities, vascular growth factor assay, and for histopathology. The assay is dependent upon the presence of integrin $\alpha v \beta 3$ in the plasma membrane of the rapidly dividing avian endothelial cell and has facilitated the identification of small molecule receptor sites, including that for thyroid hormone [12–15]. The activities of mammalian vascular growth factors, such as vascular endothelial growth factor (VEGF), basic fibroblast growth factor (bFGF or FGF2), are readily demonstrated in the CAM (see Fig. 1.2). The embryo in each CAM assay sample is subject to invasion from tumor xenografts and gives an insight into tumor aggressiveness in the presence and absence of test materials placed on the filter discs. Evaluation of angiogenesis inhibitors can be assessed using the CAM model (see Fig. 1.3).

Basal angiogenesis is generous in the CAM; pro-angiogenic test materials with low-potency effects on neovascularization may be difficult to evaluate in this system. The model yields information about the rate of onset of effects of test substances, but their decay pharmacokinetics are inferred only after removing the filter discs containing the test substances from the CAM and monitoring vessel formation or regression. Although the CAM is relatively immunotolerant, there are occasional inflammatory responses encountered about xenografts. The ability of the membrane to express inflammation, however, can be exploited when a model of the inflammatory response is desired.

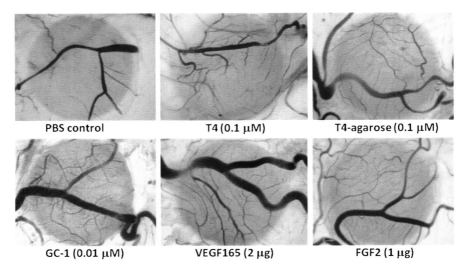

Fig. 1.2 Representative images of effects of inducers of angiogenesis in the CAM model. T4, T4-agarose and GC-1 are the thyroid hormone-based pro-angiogenic compounds [12–14]

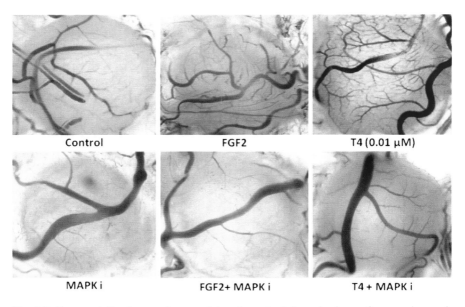

Fig. 1.3 Representative images for examining the potential mechanisms of pro-angiogenesis mediators in the CAM model. The MAPK inhibitor (MAPK i) in these studies was PD98059 [15]

Zebrafish

The zebrafish is a freshwater fish of the tropics whose embryos are relatively easy to cultivate in the laboratory. Hundreds of embryos may be generated by a mating. They develop outside the mother and are yolk sac-dependent. The embryos are optically clear and thus organ development and blood vessel formation are easily monitored by microscope [16]. A variety of optical techniques have been used in the past to visualize vessels, including confocal microscopy/angiography. Transgenic zebrafish are now readily available that express green fluorescent protein (GFP) variants that facilitate vessel visualization. Another transgenic model offers red blood cells that are labeled with a second fluorescent dye that enables blood flow to be quantitated as blood vessel formation is monitored by GFP.

The zebrafish model is one focused on embryo development and thus it may be important to distinguish between vasculogenesis and angiogenesis in studies of pharmacologic agents or growth factors. This distinction can be made anatomically, e.g., the intersegment blood vessels are thought to reflect angiogenesis in which endothelial cells develop from primordial vessel cells prior to sprouting, whereas the dorsal aorta is a component of embryonic vasculogenesis in which this differentiation does not occur. A very useful asset of this model is that it is suitable for specific gene knockdowns with siRNAs to relatively quickly assess function of genes of particular interest.

It is not yet clear how homologous fish circulation and fish angiogenesis are to human tumor-associated blood vessel formation. Another concern is the route of administration required for testing of specific pharmaceuticals or biological products. Proteins, for example, must be administered into the yolk sac. In contrast, small lipophilic compounds added to the fishtank water will be absorbed by the embryos.

Subcutaneous and Orthotopic Xenografts

The desirability of having orthotopic tumor implants (xenografts) where the local microenvironment is more homologous to that in human subjects is obvious from the standpoint of tumor biology [5]. Although still imperfect, subcutaneous tumor implants are a technically convenient option for screening numbers of compounds in dose escalation. The yield of vascular tissue about the tumors is high for histologic and biochemical analysis, simplified quantitation of vascularity (for example, by tumor hemoglobin content) is feasible, and the technical requirements of orthotopy are avoided. Many of the xenografts can be established in 2 weeks, followed by a comparable period of treatment. Subcutaneous tumor volumes may be noninvasively estimated daily, and subsets of animals can be sacrificed at intervals during treatment to obtain required histologic and biochemical information. The time period for each set of assays is not excessive relative to the amount of information that is acquired about numbers of compound(s) or combinations to be tested and dose estimation (see Fig. 1.3). Thus, from the standpoint of angiogenesis (vs. tumor cell biology), subcutaneous implants remain a useful screening option in evaluation of anti-angiogenesis agents.

Tumor measurements length (L), width (W) and height (H) can be obtained when palpable tumors are present, usually within 7 days after tumor inoculation depending on the cancer cell type. Tumor-associated angiogenesis can be determined with immunostaining using CD31 or other markers of endothelial cells or by the spectrophotometric method for hemoglobin using Drabkin's Reagent.

Compared to subcutaneous implantations, orthotopic human tumor implants in the nude mouse much more closely mimic the local environmental conditions of the primary tumor in the human subject. The behavior of tumors and associated vasculature is the result of the interaction of intrinsic qualities of the cancer cells and local host factors [17]. The interstitial pressure achieved in orthotopically implanted tumors may be quite different from that achieved when implanted elsewhere, and a variety of blood vessel properties are unique to the orthotopic site. These properties include microvascular density, permeability of vessels and vessel cell gene expression [18]. Depending upon the human tumor tissue type, such transcriptional differences among implantation sites have involved tumor genes for *FGF2* (*bFGF*), *VEGF* and VEGF receptors and, importantly, genes for multidrug resistance proteins and inflammation-relevant proteins (see review by M. Loi [5]).

The small size of the nude mouse permits repeated non-radioisotopic fluorescence IVIS imaging of orthotopic cancer implant volumes (see Fig. 1.4; [19]) that is more accurate than noninvasive tumor volume measurements of subcutaneous tumors. The IVIS scans also allow distinctions between viable cells and dead cells that contribute to tumor volume and can measure other properties of the cells in the scanned lesion.

Matrigel® Plug Assay

Both *in vitro* and *in vivo* methods have been described that utilize Matrigel® or collagen or fibrin as a matrix in which differentiation of endothelial cells may be monitored in response to vascular growth factors, to other pro-angiogenic factors and to anti-angiogenic agents. Differentiation is vessel sprouting or microtubule formation, depending upon material selected to form the matrix. In Matrigel® it is tubule formation that results. Alternatively, Matrigel® plugs to which a vascular growth factor has been added will cause host blood vessel-based angiogenesis in the plug, and this process can be quantitated. Matrigel® is composed of mouse sarcoma extracellular matrix proteins and basement membrane proteins. It is a solid gel at 37 °C and liquid at 4 °C. Matrigel® contains murine vascular growth factors harvested with extracellular matrix from the original mouse sarcoma. A preparation of Matrigel® is available from which some of the mouse pro-angiogenic substances have been removed.

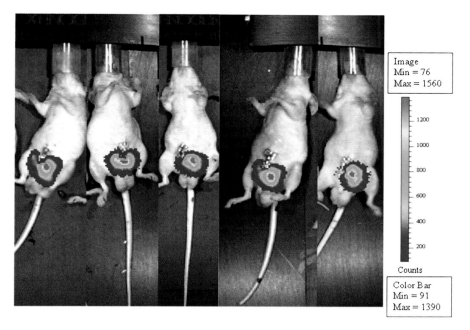

Fig. 1.4 Representative IVIS images taken 1 week after orthotopic implantation of prostate cancer cells (PC3-Luc) in male nude mice. Loss of cell viability is tracked by bioluminescent color change (red (actively dividing cells) to blue)

An *in vivo* protocol we have used [11, 20] induces host vascularization of plugs over 7–14 days. One or more subcutaneous injections per mouse are done of standard volumes of Matrigel®, e.g., 100 µL, containing a vascular growth factor such as FGF2. The anti-angiogenesis compound is administered systemically. A rapidly measured index of drug action on blood vessel formation is hemoglobin content of standard volumes of harvested aliquots of plugs, but semi-quantitative histologic estimates of angiogenesis may also be applied. This approach requires knowledge of the pharmacokinetics of the test substance(s). The duration of the assay is comparable to that of xenograft models.

In Vitro Assays

Aortic Ring Assay

The standard mouse aortic ring assay requires harvest of the thoracic aorta, excision of the adventitia and serial cutting of rings 1 mm in length that are implanted in collagen gels [21, 22]. Culture in serum-free medium for 7 days results in new vessel outgrowth that can be evaluated by microscopy and immunohistochemical staining and compared with that in rings exposed to the agent(s) of interest. The mouse aortic ring assay offers the substantial advantage of studying transgenic animals in which factors that contribute to angiogenesis are present, although the rat aorta has also been used and has a larger ring yield.

This is a somewhat difficult assay technically. Factors such as incomplete adventitia excision can influence the vessel outgrowth, as may strain and age of the mouse source of tissue (see review by A.M. Goodwin [1]).

Sprouting Assay

A variety of *in vitro* methods have been developed that take advantage of vessel sprouting that can be induced by vascular growth factors in cultures of endothelial cells. Sprouting can be encouraged in confluent monolayers of endothelial cells on fibrin gels or the cells can be grown to confluence on microbeads that are then suspended in fibrin gels and exposed to test substances. Sprouting is quantitated microscopically. In the case of the microbead assay, sprouting from a bead may be defined in terms of vessel length or numbers of vessels/bead.

A system we have exploited involves human dermal microvascular endothelial cells (HDMECs) grown to confluence in several days on Cytodex-3 beads that are gelatin-coated, exposed to normal human serum and placed in fibrin that undergoes polymerization [23]. The system is ready for addition of pro- or anti-angiogenic agents within 48 h. In this model we have demonstrated the anti-angiogenic activity— inhibition of actions of FGF2 and VEGF—of a thyroid hormone derivative, tetraiodothyroacetic acid (tetrac) [11], whose receptor site is on integrin $\alpha v \beta 3$ on the surface of rapidly dividing endothelial cells [12].

Tube Formation Assay

Angiogenesis depends upon vessel tube formation from endothelial cells [24, 25]. Tube formation can be modeled by incubating plate-cultured endothelial cells with extracellular matrix components, such as those found in Matrigel®. Cells used in this system are human umbilical vein endothelial cells (HUVECs) or HDMECs plated on Matrigel®. Quantitation of angiogenic or anti-angiogenic activities is by analysis of photographs of plate wells for length and area of capillary-like structures (CLS)/unit area (Fig. 1.5). Assay duration is short, as little as 1 day.

The murine vascular growth factors in Matrigel® can complicate interpretation of results obtained in this model. It is essential that the integrity of the CLS be assured by the presence of lumens in the vessels, since an endothelial cell may extend bridges or cords to other cells. Fibroblasts and other cells may form networks on Matrigel® [26], and thus the cultured endothelial cells must be free of contamination with fibroblasts or cancer cells.

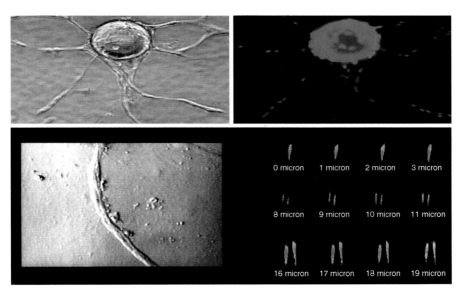

Fig. 1.5 Representative images of tube formation using three dimensional tube formation assay

Co-culture of Human Cancer Cells with Human Microvascular Endothelial Cells

A two-chamber co-culture model for studying endothelial cell sprouting and capillary formation has been utilized to study human lung carcinoma [27, 28] or murine breast carcinoma-induced angiogenesis [29, 30]. In this model, quantifiable capillaries form by sprouting from a confluent EC monolayer after co-culture with cancer cells is initiated. This co-culture system has been adapted for the study of capillary formation in an *in vitro* human mammary carcinoma model. A schematic diagram and description of the co-culture model is shown in Fig. 1.6. Each well contains a Transwell insert (T), with a polycarbonate membrane (M), 10 μm thick with 0.8 μm pores. Insert (T) delineates upper chamber (U) and lower chamber (L) with the levels of media equal in both chambers. Figure 1.6b represents a magnification of the separation of the chambers; the top arrow points to HDMECs seeded in the upper chamber on reduced growth factor (RGF)-Matrigel-coated surface to simulate basement membrane that underlies the endothelium *in vivo*, and the bottom arrow points to tumor cells (MCF-7, MDA-MB) seeded on type I collagen. Individual cell populations can be manipulated separately up until the time of co-culture. Test reagents can be placed in upper, lower or both compartments. Capillary formation is evaluated by staining of HDMECs using EC-specific CD31/PECAM antibody followed by quantification of digital images as previously described [27].

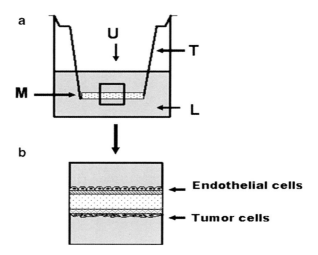

Fig. 1.6 A schematic diagram and description of the co-culture model. (**a**) Each well contains a Transwell insert (T), with a polycarbonate membrane (M). Insert (T) delineates upper chamber (U) from lower chamber (L) with the levels of media equal in both chambers. (**b**) Represents magnification of the separations between both chambers. *Top arrow* points to endothelial cells seeded in the upper chamber on reduced growth factor (RGF)-Matrigel-coated surface and *bottom arrow* points to cancer cells seeded on type I collagen

Conclusions

An idealized angiogenesis assay would be a skin window in a subhuman primate by which angiogenesis-relevant agents may be directly applied and effects monitored microscopically following systemic administration. Such an assay would be quantitative, e.g., subject to monitoring of the number of angiogenic buds/unit area via a computer-imposed template. They would permit local sampling of concentrations of host factors, and biopsy must be feasible to allow histologic analysis, including serial sampling/biopsy. This idealized assay of course does not currently exist and would be valid for studies of wound-healing-related angiogenesis, but not for neovascularization associated with various tumors, where orthotopic models are best.

We have reviewed here a complement of widely-used assays. Each has validated utility and has provided information useful to the evaluation and development of specific pro- or anti-angiogenic pharmaceuticals or to the activities of naturally occurring factors that modulate new blood vessel formation. Each assay has defects, as we point out. Strategically, several assays will be used in the evaluation of compounds. One should be a cost-effective screening method that permits initial estimations of dosing, comparison of congeners, comparison with activities of naturally occurring angiogenic substances and combinations of agents. A second method should be used for confirmation of pro- or anti-angiogenic activity and, if its mechanisms are sufficiently understood, for initial studies of the molecular basis of action. Intact animal models may then be used for validation of effectiveness in ischemia-reperfusion injury, wound-healing and disruption of vascular support for cancers. Effectiveness in certain of these will require quantitative and qualitative histopathology, vascular radiology and high resolution *in vivo* imaging systems [31]. An example of quantitative histopathology is the number of small vessels and capillaries. The *in vivo* imaging systems include fluorescent dyes and micro-computerized tomography.

It may also be useful to quantitate the motility response of endothelial cell or fibroblasts to the angiogenic agents under evaluation. The standard technique is use of the Boyden chamber apparatus, in which cells migrate through a porous membrane from an upper to a lower chamber where the latter contains a biochemical cue, such as vitronectin [32].

From the standpoint of more fully elucidating the mechanisms by which angiogenesis-relevant pharmaceuticals work, it is desirable to have models in which specific components of the angiogenesis process, e.g., signal transducing kinases or other enzymes, are reduced (knocked-down) in target blood vessel cells by small interfering RNAs (siRNAs). The siRNA approach may be used in several angiogenesis assay systems and the *in vitro* endothelial cell-fibroblast organotypic co-culture system is an attractive model for this modification [29].

References

1. Goodwin AM (2007) In vitro assays of angiogenesis for assessment of angiogenic and anti-angiogenic agents. Microvasc Res 74(2–3):172–183
2. Ucuzian AA, Greisler HP (2007) In vitro models of angiogenesis. World J Surg 31(4):654–663
3. Staton CA, Reed MWR, Brown NJ (2009) A critical analysis of current *in vitro* and *in vivo* angiogenesis assays. Int J Exp Path 90:195–221
4. Jensen LD, Cao R, Cao Y (2009) In vivo angiogenesis and lymphangiogenesis models. Curr Mol Med 9(8):982–991
5. Loi M, Di Paolo D, Becherini P, Zorzoli A, Perri P, Carosio R, Cilli M, Ribatti D, Brignole C, Pagnan G, Ponzoni M, Pastorino F (2011) The use of the orthotopic model to validate antivascular therapies for cancer. Int J Dev Biol 55:547–555
6. Cimpean A-C, Ribatti D, Raica M (2011) A brief history of angiogenesis assays. Int J Dev Biol 55:377–382
7. Auerbach R, Kubai L, Knighton D, Folkman J (1974) A simple procedure for the long-term cultivation of chicken embryos. Dev Biol 41(2):391–394
8. Ribatti D, Conconi MT, Nussdorfer GG (2007) Nonclassic endogenous novel regulators of angiogenesis. Pharmacol Rev 59(2):185–205
9. Vargas A, Zeisser-Labouebe M, Lnge N, Gurny R, Delie F (2007) The chick embryo and its chorioallantoic membrane (CAM) for the in vivo evaluation of drug delivery systems. Adv Drug Deliv Rev 59(11):1162–1176
10. Mousa SS, Mousa SA (2005) Effect of resveratrol on angiogenesis and platelet/fibrin-accelerated tumor growth in the chick chorioallantoic membrane model. Nutr Cancer 52(1):59–65
11. Mousa SA, Bergh JJ, Dier E, Rebbaa A, O'Connor LJ, Yalcin M, Aljada A, Duskin E, Davis FB, Lin HY, Davis PJ (2008) Tetraiodothyroacetic acid, a small molecule integrin ligand, blocks angiogenesis induced by vascular endothelial growth factor and basic fibroblast growth factor. Angiogenesis 11(2):183–190

12. Bergh JJ, Lin HY, Lansing L, Mohamed SN, Davis FB, Mousa S, Davis PJ (2005) Integrin αvβ3 contains a cell surface receptor site for thyroid hormone that is linked to activation of mitogen-activated protein kinase and induction of angiogenesis. Endocrinology 146:2864–2871

13. Davis PJ, Davis FB, Mousa SA (2009) Thyroid hormone-induced angiogenesis. Curr Cardiol Rev 5:12–16

14. Davis PJ, Davis FB, Mousa SA, Luidens MK, Lin HY (2011) Membrane receptor for thyroid hormone: physiologic and pharmacologic implications. Annu Rev Pharmacol Toxicol 51:99–115

15. Lin HY, Davis FB, Gordinier JK, Martino LJ, Davis PJ (1999) Thyroid hormone induces activation of mitogen-activated protein kinase in cultured cells. Am J Physiol 276:C1014–C1024

16. Tobia C, De Sena G, Presta M (2011) Zebrafish embryo, a tool to study tumor angiogenesis. Int J Dev Biol 55(4–5):505–509

17. Talmadge JE, Donkor M, Scholar E (2007) Inflammatory cell infiltration of tumors: Jekyll or Hyde. Cancer Metastasis Rev 26(3–4):373–400

18. Langenkamp E, Molema G (2009) Microvascular endothelial cell heterogeneity: general concepts and pharmacological consequences for anti-angiogenic therapy of cancer. Cell Tissue Res 335(1):205–222

19. Yalcin M, Lin HY, Sudha T, Bharali DJ, Meng R, Tang HY, Davis FB, Stain SC, Davis PJ, Mousa SA (2013) Response of human pancreatic cancer cell xenografts to tetraiodothyroacetic acid nanoparticles. Horm Cancer. doi:10.1007/s12672-0137-y

20. Powell JA, Mohamed SN, Kerr JS, Mousa SA (2000) Antiangiogenesis efficacy of nitric oxide donors. J Cell Biochem 80(1):104–114

21. Nicosia RF, Otinetti A (1990) Modulation of microvascular growth and morphogenesis by reconstituted basement membrane gel in three-dimensional cultures of rat aorta: a comparative study of angiogenesis in matrigel, collagen, fibrin and plasma clot. In Vitro Cell Dev Biol 26(2):119–128

22. Zhu WH, Nicosia RF (2002) The thin prep rat aortic ring assay: a modified method for the characterization of angiogenesis in whole mounts. Angiogenesis 5(1–2):81–86

23. Mousa SA, O'Connor L, Davis FB, Davis PJ (2006) Proangiogenesis action of the thyroid hormone analog 3, 5-diiodothyropropionic acid (DITPA) is initiated at the cell surface and is integrin mediated. Endocrinology 147(4):1602–1607

24. Kubota Y, Kleinman HK, Martin GR, Lawley TJ (1988) Role of laminin and basement membrane in the morphological differentiation of human endothelial cells into capillary-like structures. J Cell Biol 107(4):1589–1598

25. Arnaoutva I, George J, Kleinman HK, Benton G (2009) The endothelial cell tube formation assay on basement membrane turns 20: state of the science and the art. Angiogenesis 12(3):267–274

26. Donovan D, Brown NJ, Bishop ET, Lewis CE (2001) Comparison of three in vitro human 'angiogenesis' assays with capillaries formed in vivo. Angiogenesis 4(2):113–121

27. Phillips PG, Birnby LM, Narendran A, Milonovich WL (2001) Nitric oxide modulates capillary formation at the endothelial cell-tumor cell interface. Am J Physiol 281:278–290

28. Phillips PG, Birnby L (2004) Nitric oxide modulates Caveolin-1 (Cav-1) and MMP-9 expression and distribution at the endothelial cell/tumor cell interface. Am J Physiol Lung Cell Mol Physiol 286:L1055–1065

29. Lincoln DW II, Phillips PG, Bove K (2003) Estrogen-induced Ets-1 promotes capillary formation in an in vitro tumor angiogenesis model. Breast Cancer Res Treat 78:167–178

30. Hetheridge C, Mavria G, Mellor H (2011) Uses of the in vigor endothelial-fibroblast organotypic co-culture assay in angiogenesis research. Biochem Soc Trans 39(6):1597–1600

31. Wessels JT, Busse AC, Mahrt J, Dulin C, Grabbe E, Mueller GA (2007) In vivo imaging in experimental preclinical tumor research—a review. Cytometry A 71(8):542–549

32. Colman RW, Jameson BA, Lin Y, Johnson D, Mousa SA (2000) Domain 5 of high molecular weight kininogen (kininostatin) downregulates endothelial cell proliferation and migration and inhibits angiogenesis. Blood 95(2):543–560

Part I
Pro-angiogenesis Targets and Clinical Implications

Chapter 2
Survey of Pro-angiogenesis Strategies

Shaker A. Mousa

Abstract Angiogenesis, the formation of new capillaries from pre-existing vascular network, plays an important role in physiological processes such as embryonic development, wound-healing, and pathological processes such as cancer, ocular neovascularization, and certain inflammatory-associated disorders. Even though pro-angiogenesis has been shown to be safe and well-tolerated in clinical trials, efficacy of the treatment has not been satisfactory. Systemic use of pro-angiogenesis factors might accelerate pathological states by inducing microvessel growth in, for example, tumors and atherosclerotic lesions. Hence, stimulation of angiogenesis is preferred to be used locally at the desired site or tissues.

The key signaling system that regulates proliferation and migration of endothelial cells forming the basis of any vessel are vascular endothelium growth factors (VEGF) and their receptors. A number of other signaling systems are also involved in regulation of the main steps of vessel formation. The signaling system Dll4/Notch regulates selection of endothelial cells for beginning of angiogenesis expansion. An important step in vessel stabilization and maturation is vascular wall formation, which might be stimulated by certain glycosaminoglycan [1]. Signaling system PDGFB/PDGFR beta as well as angiopoietins Ang-1, Ang-2, and their receptor Tie2 besides other pathways are involved in recruiting pericytes and smooth muscle cells. See Table 2.1 for pro-angiogenesis factors and mediators.

The goal of therapeutic angiogenesis is to improve perfusion and restore tissue function, leading to a broad range of interventions that generate new blood vessel growth to promote neovascularization for wound-healing, diabetic ulcers, peripheral arterial disease, and tissue repair including critical limb ischemia and ischemic

S.A. Mousa (✉)
The Pharmaceutical Research Institute at Albany College of Pharmacy
and Health Sciences, Rensselaer, NY, USA
e-mail: shaker.mousa@acphs.edu

S.A. Mousa and P.J. Davis (eds.), *Angiogenesis Modulations in Health and Disease:*   15
Practical Applications of Pro- and Anti-angiogenesis Targets,
DOI 10.1007/978-94-007-6467-5_2, © Springer Science+Business Media Dordrecht 2013

Table 2.1 Pro-angiogenesis
factors/mediators

Pro-angiogenesis factors
Vascular endothelial growth factor (VEGF)
Fibroblast growth factor (FGF)
Hepatocyte growth factor (HGF)
Platelet derived growth factor (PDGF)
Epidermal growth factor (EGF)
Placental growth factor (PlGF)
Cytokines (TNF-α, IL-8,…)
Chemokines
Matrix metalloproteinases (MMPs)
Complement activators
Hypoxia inducible factor-1α (HIF-1α)
Angiopoietin
Angiogenin
Tissue factor/Factor VIIa
Thrombin
Protease activated receptor (PAR)
Certain glycosaminoglycan fragments
Thyroid agonists (L-T3, L-T4,…)
Nicotinic receptor agonists
Oxidative stress (H_2O_2, Lipid hydroperoxide)
Certain lipid-derived (LPA, S1P,…)

heart disease. Although experimental studies on the stimulation of arteriogenesis have been promising, not a single drug has been proved to be applicable in clinical practice, either because of lack of efficacy or because of undesired side effects.

Thyroid hormone agonists and analogues demonstrated potent pro-angiogenesis efficacy, with the potential utility in therapeutic angiogenesis [2–5] as reviewed in subsequent chapters. Other hormones play a role in angiogenesis modulation as reviewed in a subsequent chapter.

The neurotransmitter/receptor system such as nicotine (nicotinic receptor) and catecholamine has been shown to modulate various aspects of angiogenesis processes as reviewed in the subsequent chapters. It has also been found that tumor tissues can not only synthesize and release a wide range of neurotransmitters, but can also produce different biological effects via respective receptors. Neurotransmitters can produce either stimulatory or inhibitory effects in normal and tumor tissues. These effects are dependent on the types of tissues and the kinds of neurotransmitter as well as the subtypes of corresponding receptors involved. These findings clearly extend the conventional role of neurotransmitters in the nervous system to the actions in oncogenesis and other angiogenesis-associated disorders. In this regard, intervention or stimulation of these neuronal pathways in different cancer diseases would have significant clinical implications in angiogenesis-associated processes [6, 7].

Certain arachidonic acid-derived lipid mediators have positive regulation of angiogenesis while others have negative modulation of angiogenesis, and the balance between these mediators might be shifted in pathological conditions associated with accelerated angiogenesis and vice versa. Lysophosphatidic acid (LPA) and sphingosine 1-phosphate (S1P) are naturally arising bioactive lipids. The roles of LPA and S1P in angiogenesis, tumor growth, and metastasis have recently emerged. Platelets, beside other cells, are an important source of LPA and S1P. LPA and S1P interact with a large series of G-protein-coupled receptors, which may account for the wide variety of cell types reacting to LPA and S1P. These lipid-derived pro- or anti-angiogenesis mediators along with their receptors represent novel pharmacological targets in the modulation of angiogenesis.

Peroxisome proliferator-activated receptors (PPARs) are a group of nuclear hormone receptors that regulate lipid and glucose metabolism. PPAR-α agonists such as fenofibrate and PPAR-γ agonists such as the thiozolidinediones have been used to treat dyslipidemia and insulin resistance in diabetes. Over the past few years the role of PPARs in the regulation of angiogenesis has been discovered [8, 9].

Therapeutic Angiogenesis Targets

Growth factor-based therapies include the only FDA-approved recombinant protein drug rhPDGF (becaplermin, REGRANEX® 0.01 % gel) indicated for diabetic neuropathic lower extremity ulcers. Growth factors delivered through autologous isolates of patient platelets include AutoloGel®, and SmartPReP®. Currently, there are no FDA-approved pro-angiogenesis drugs for the treatment of ischemic cardiovascular disease.

Tissue Engineering and Cell-based Therapies

Stimulation of neovascularization is critical for the success of tissue engineering. Scaffold and extracellular matrix surfaces are expected to play a major role in the promotion of neovascularization and the support of tissues to be engineered. Tissue engineered products approved by the FDA include the bilayered skin substitute Grafstkin (Apligraf®) and the fibroblast dermal skin substitute Dermagraft®. These products contain living or cryopreserved cells on a matrix capable of secreting and releasing multiple pro-angiogenesis growth factors into the wound bed. Additionally, CD34+ endothelial progenitor cells (EPC) derived from bone marrow or from peripheral blood have been found to enhance angiogenesis in ischemic tissues [10, 11], increase transcutaneous oxygen, and increase collateral vessels. Additionally, trophic factors secreted by EPC might have an impact on therapeutic angiogenesis [12].

References

1. Mousa SA, Feng X, Xie J, Du Y, Hua Y, He H, O'Connor L, Linhardt RJ (2006) Synthetic oligosaccharide stimulates and stabilizes angiogenesis: structure-function relationships and potential mechanisms. J Cardiovasc Pharmacol 48(2):6–13
2. Davis PJ, Davis FB, Mousa SA (2009) Thyroid hormone-induced angiogenesis. Curr Cardiol Rev 5(1):12–16
3. Mousa SA, Davis FB, Mohamed S, Davis PJ, Feng X (2006) Pro-angiogenesis action of thyroid hormone and analogs in a three-dimensional in vitro microvascular endothelial sprouting model. Int Angiol 25(4):407–413
4. Mousa SA, O'Connor LJ, Bergh JJ, Davis FB, Scanlan TS, Davis PJ (2005) The proangiogenic action of thyroid hormone analogue GC-1 is initiated at an integrin. J Cardiovasc Pharmacol 46(3):356–360
5. Davis FB, Mousa SA, O'Connor L, Mohamed S, Lin HY, Cao HJ, Davis PJ (2004) Proangiogenic action of thyroid hormone is fibroblast growth factor-dependent and is initiated at the cell surface. Circ Res 94(11):1500–1506
6. Li ZJ, Cho CH (2011) Neurotransmitters, more than meets the eye–neurotransmitters and their perspectives in cancer development and therapy. Eur J Pharmacol 667(1–3):17–22
7. Mousa S, Mousa SA (2006) Cellular and molecular mechanisms of nicotine's pro-angiogenesis activity and its potential impact on cancer. J Cell Biochem 97(6):1370–1378
8. Aljada A, O'Connor L, Fu YY, Mousa SA (2008) PPAR gamma ligands, rosiglitazone and pioglitazone, inhibit bFGF- and VEGF-mediated angiogenesis. Angiogenesis 11:361–367
9. Biscetti F, Gaetani E, Flex A et al (2008) Selective activation of peroxisome proliferator-activated receptor (PPAR)alpha and PPAR gamma induces neoangiogenesis through a vascular endothelial growth factor-dependent mechanism. Diabetes 57:1394–1404
10. Chen YL, Chang CL, Sun CK, Wu CJ, Tsai TH, Chung SY, Chua S, Yeh KH, Leu S, Sheu JJ, Lee FY, Yen CH, Yip HK (2012) Impact of obesity control on circulating level of endothelial progenitor cells and angiogenesis in response to ischemic stimulation. J Transl Med 10:86. doi:10.1186/1479-5876-10-86
11. Fan Y, Shen F, Frenzel T, Zhu W, Ye J, Liu J, Chen Y, Su H, Young WL, Yang GY (2010) Endothelial progenitor cell transplantation improves long-term stroke outcome in mice. Ann Neurol 67(4):488–497
12. Di Santo S, Yang Z, Wyler von Ballmoos M, Voelzmann J, Diehm N, Baumgartner I, Kalka C (2009) Novel cell-free strategy for therapeutic angiogenesis: in vitro generated conditioned medium can replace progenitor cell transplantation. PLoS One 4(5):e5643

Chapter 3
Angiogenesis Modulation by Arachidonic Acid-derived Lipids: Positive and Negative Regulators of Angiogenesis

Robert C. Block, Murat Yalcin, Mathangi Srinivasan, Steve Georas, and Shaker A. Mousa

Abstract Arachidonic acid-derived lipids such as 15 deoxy-PGJ2 or 15 epi-lipoxin A4 have been shown to be potent anti-angiogenesis agents regardless of the angiogenesis stimulus. Other arachidonic acid-derived mediators differentially stimulate angiogenesis and the balance among the different arachidonic acid metabolites along their interactions might play an important role in angiogenesis hemostasis in various angiogenesis-associated disorders.

R.C. Block (✉)
Department of Public Health Sciences, University of Rochester School
of Medicine and Dentistry, Rochester, NY, USA

Division of Cardiology, Department of Medicine, University of Rochester School of Medicine
and Dentistry, Rochester, NY, USA
e-mail: Robert_Block@URMC.Rochester.edu

M. Yalcin
The Pharmaceutical Research Institute at Albany College of Pharmacy
and Health Sciences, Rensselaer, NY, USA

Veterinary Medicine Faculty, Department of Physiology, Uludag University, Gorukle,
Bursa, Turkey

M. Srinivasan • S.A. Mousa
The Pharmaceutical Research Institute at Albany College of Pharmacy
and Health Sciences, Rensselaer, NY, USA

S. Georas
Division of Pulmonary and Critical Care Medicine, Department of Medicine,
University of Rochester School of Medicine and Dentistry,
Rochester, NY, USA

S.A. Mousa and P.J. Davis (eds.), *Angiogenesis Modulations in Health and Disease:*
Practical Applications of Pro- and Anti-angiogenesis Targets,
DOI 10.1007/978-94-007-6467-5_3, © Springer Science+Business Media Dordrecht 2013

Introduction

Cardiovascular disease (CVD) has been established as the leading cause of death of adults in the United States [1], and is now the leading cause of death globally [2]. Although classic risk factors for CVD are well characterized and tools for assessing and altering risk are readily available [1, 3], sudden cardiac death remains the cause of approximately 63 % of cardiac mortality [4]. Atherosclerosis is the primary etiology of sudden cardiac death, and it has been estimated that 69 % of women who suffer sudden cardiac death had no prior history of heart disease [5]. The burden from this disease is so great that deaths from all cancers combined would need to be summed in order to exceed that from sudden cardiac death [6]. Although atherosclerotic plaques are the primary instigator of myocardial ischemia, it is the vulnerable and unstable plaque that is the cause of the majority of myocardial infarctions and cardiac death [7].

Following CVD, cancer is the second leading cause of death in the United States [8, 9]. Despite substantial progress in the therapeutic options available for treating cancer, including therapeutics for some of the most common solid neoplasms (breast, colon, and lung) [10], mortality from cancer remains a major public health challenge.

The present investigation examined the interactions among various AA-derived or containing lipid products in the modulation of angiogenesis that might have an impact in CVD and cancer. We studied the effects of several AA-derived lipid mediators in comparison to known pro-angiogenesis factors on angiogenesis in an effort to elucidate how shifts in lipid metabolism and the interaction of various lipid metabolites with each other, perhaps due to innate differences among individuals and/or medication use, can alter the hemostatic and angiogenic balance and potentially regulate the risk for cardiovascular events, neoplasia, and the pathophysiology of other diseases.

Angiogenesis

Angiogenesis is the process whereby new capillary networks are formed from preexisting blood vessels by the sprouting and/or splitting of capillaries [11]. This process involves the orchestrated proliferation, migration, adhesion, differentiation, and assembly of vascular endothelial cells and their surrounding vascular smooth muscle cells. Angiogenesis is an integral component of physiological processes, including wound-healing and myocardial angiogenesis after ischemic injury. It is precisely controlled by the balance of pro-angiogenic factors such as VEGF, bFGF, and platelet-derived growth factor, and anti-angiogenic factors. Disruption of the normal, regulated process of angiogenesis can result in pathological conditions such as solid tumor growth, atherosclerosis, hypertension, rheumatoid arthritis, and diabetic retinopathy.

Angiogenesis is a complex process involved in the pathophysiology of cardiovascular diseases and cancer. Although the role of angiogenesis in the process of atherosclerosis is controversial, the discovery of an association between intimal neovascularization and atherosclerosis dates to the year 1876, when it was noted by Koester [12]. The development of vasa vasorum through angiogenesis has been associated with plaque instability and rupture, as microvessel formation has a predilection for the shoulder regions of atherosclerotic plaques [12–14]. Therapeutic modulators of angiogenesis that would induce the growth of new blood vessels and thus potentially reduce ischemic burden in the heart and limbs were greatly anticipated [15], however, despite promising results in preclinical models, data from clinical trials have been inconclusive, and evidence suggests that angiogenic factors actually promote atherosclerosis and potentially destabilize coronary plaques [16, 17]. The role of angiogenesis in solid cancer growth is less controversial and efforts to develop anti-angiogenic therapeutics have been more successful.

Arachidonic Acid

Arachidonic acid (AA) is an omega-6 long-chain, polyunsaturated fatty acid that in humans is primarily derived from diet. It has been shown to play a crucial role in the maintenance of homeostasis [18, 19]. AA is the source of thromboxane, a highly robust platelet function activator [20], as well as a number of classic pro-inflammatory prostaglandins and leukotrienes produced via metabolism by the cyclo-oxygenase and lipoxygenase pathways [21]. AA is also the source of other, more recently discovered lipid mediators, including epoxyeicosatrienoic acids (EETs), which are produced by the liver via the cytochrome P450 enzyme pathway and have anti-inflammatory and anti-atherosclerotic activity [22, 23]. At the same time, there is evidence that EETs have a detrimental role through the stimulation of myocardial fibrosis [24]. Lysophosphatidylcholine (LPC) and lysophosphatidic acid (LPA) are produced via the actions of phospholipases and other enzymes [25]. Derived from AA are the families of lipoxins, some of which are triggered by aspirin, and 15 deoxy-prostaglandin D2, the ligand for peroxisome proliferator-activated receptor (PPAR)-γ. These lipid mediators have been shown to exert potent anti-inflammatory effects [26].

Effects of AA-derived Metabolites/Lipid Mediators on Angiogenesis

We investigated the differential effects of AA-derived metabolites/lipid mediators on angiogenesis using a well-validated *in vivo* CAM model system. The data revealed that LPA, LPC, and the EETs are pro-angiogenic, while 15 deoxy-PGJ2 and 15 epi-lipoxin A4 inhibit angiogenesis-mediated by AA-derived and non-AA derived pro-angiogenesis factors. In addition, these results suggest that interactive effects among different lipid mediators may differentially regulate angiogenesis. All of the lipid mediators (LPA, EETs, 15 deoxy-PGJ2, and 15 epi-lipoxin A4) that we examined are derived from AA, a fatty acid that is reliably present in human blood due to diet [27, 28], but is also produced via varying metabolic pathways, including the de-saturation and elongation of the 18-carbon omega-6, linoleic acid. Although the effects of these endogenous lipid mediators on hemostasis and angiogenesis have been previously documented, the focus has traditionally been on the actions of individual compounds. However, the translation of such data to effects in humans is complicated by many factors that influence metabolism, including genetics, medications, and lifestyle. The individual effects of any cell signaling or modulatory molecule will be influenced by other, sometimes antagonistic, molecules that compete for the same metabolic pathways or binding sites. Thus, the effects driven by any molecule in isolation may not reflect the situation *in vivo*. That is, the observed effects of a single molecule alone may be significantly altered *in vivo* due to interactive or negative effects from other molecules or factors. The mediators examined in the current study have specific mechanisms of action that influence underlying pathways that intersect with atherosclerotic and angiogenic processes [29–32]. As such, these mediators and their mechanisms of action can impinge on drugs that are commonly used in the treatment and prevention of cardiovascular diseases. These drugs include aspirin, the thiazolidinediones, and drugs that impact the cytochrome P450 systems. The current data indicate that AA-derived lipid mediators and their associated metabolic pathways can interact in as-yet undefined ways to potentially regulate angiogenesis. These findings provide useful information for a better understanding of the regulatory checks and balances that govern human health and disease.

Epoxyeicosatrienoic Acids (EETs)

EETs are generated from AA by cytochrome P450 (CYP) epoxygenases, the expression of which is regulated by hemodynamic and pharmacological stimuli, as well as by hypoxia [33]. The activity of many EETs is initiated by binding of EET to a membrane receptor, resulting in activation of ion channels and intracellular signal transduction pathways [34]. However, EETs can also be taken up by cells and incorporated into phospholipids, whereby they can bind to cytosolic proteins and nuclear receptors, which suggests that some of the activities of EETs are mediated by direct interactions with intracellular effector systems. CYP enzymes are highly expressed in the liver and kidney, but have also been detected at lower levels in the brain, heart and vasculature. To date, the importance of EETs in vascular homeostasis has been difficult to demonstrate and therefore is likely largely underestimated because of the labile nature of EET-forming enzymes in cell culture. Each of the 4 EETs tested in this study (5, 6-, 8, 9-, 11, 12-, and 14, 15-EETs), have been shown to have pro-angiogenic properties, where they promote endothelial cell migration and the formation of capillary-like structures [29, 35]. The neovascularization effects of 5, 6- and 8, 9-EET are induced by the inhibition of EET enzymatic hydration. VEGF stimulates angiogenesis by increasing CYP2C promoter activity in endothelial cells, which results in increased intracellular levels of EETs [36]. VEGF-induced endothelial cell tube formation has been shown to be inhibited by the EET antagonist 14, 15-epoxyeicosa-5(Z)-enoicacid (14, 15-EEZE), and VEGF-induced endothelial cell sprouting can be attenuated by CYP2C antisense nucleotides. These data, published by others, suggest that CYP2C-derived EETs act as second messengers, and that preventing increases in CYP activity dampens the angiogenic response to VEGF. Overexpression of CYP2C9 has been shown to stimulate endothelial tube formation, and CYP2C9-induced generation of 11, 12-EET has been shown to increase COX-2 expression via a cAMP-dependent pathway [37].

Lysophospholipids LPA and S1P

The bioactive lysophospholipids LPA and S1P are membrane-derived lipid mediators produced from phospholipid precursors of membranes and secreted by platelets [38–40], macrophages [38, 41], and some cancer cells [38, 42] . The major sources of LPA and S1P are activated platelets, injured cells, and cells stimulated by growth factors, which suggests a potential role for these molecules in inflammation, wound-healing, and tumor formation [38]. LPA can also be generated by various metabolic pathways including synthesis from LPC by hydrolysis of the choline moiety [43]. Plasma LPC is thought to be derived from phosphatidylcholine in lipoproteins and membrane micro-vesicles by acyltransferases and phospholipases [44]. LPA is a simple molecule, of a glycerol backbone with a single acyl group at positions sn1 or sn2 and a phosphate group at position 3 that nonetheless displays robust cell signaling properties in such processes as cell proliferation, migration, and survival, as well as platelet aggregation, tumor cell invasion and angiogenesis [30, 45, 46]. LPA and S1P also regulate the migration, proliferation, and survival of endothelial cells [38, 47]. LPA and S1P exert their effects via multiple G protein-coupled receptors of the endothelial differentiation gene (Edg) family [48, 49], thereby mediating a broad spectrum of intracellular events, including increases in inositol phosphates and intracellular calcium [50], down-regulation of adenylyl cyclase [51], and activation of protein kinase C and many other signaling pathways [38, 52, 53]. The acyl chains in LPC and LPA molecules vary from saturated to highly unsaturated and from 16 to 22 carbons. Evidence indicates that different LPC and LPA acyl species can have different effects on target cells, possibly reflecting subtle differences in acyl-specificity of lysophospholipid receptors [54]. For example, unsaturated LPA species (16:1 LPA, 18:1, the unsaturated LPA found in highest concentration in human serum, and 18:2 LPA) induce de-differentiation and remodeling in vascular smooth muscle cells, but saturated species do not [54, 55]. However, as far as we are aware, our study is the first to identify that the arachidonic acid species of LPA, and LPC, are angiogenesis activators.

15 Epi-Lipoxin A4

Fifteen epi-lipoxin A4 is an aspirin-triggered, 15R enantiomeric counterpart of lipoxin A4 and a product of AA metabolism [56]. It is an endogenous mediator generated during multi-cellular interactions and displays potent immunomodulatory properties [26]. In human umbilical vein endothelial cells (HUVAC) culture, the 15 epi-lipoxin A4 analogue 15 epi-16 (paraflouro)-(phenoxy-lipoxin A(4) (ATL-1)) has been shown to inhibit angiogenesis through down-regulation and/or reduction of VEGF-stimulated proliferation/chemotaxis, metalloproteinase-9 activity and expression, stress fiber formation, and formation of the actin cytoskeleton [57]. ATL-1 also inhibits endothelial cell adhesion to fibronectin and extracellular matrix/collagen cross-linking through its interaction with the G-protein-linked lipoxin A4 receptor [31]. These data suggest that aspirin-triggered production of 15 epi-lipoxin A4 may modulate angiogenesis by inhibiting endothelial cell migration, through down-regulation of actin polymerization and inhibition of focal contacts. In an *in vivo* model of inflammatory angiogenesis, ATL-1 down-regulates angiogenesis by approximately 50 %. In addition, both aspirin-triggered 15 epi-lipoxin A4 and its native enantiomere lipoxin A4 display potent anti-inflammatory activities *in vivo* that are receptor-, cell-type-, and tissue-specific [57]. Aspirin-triggered 15 epi-lipoxin A4 and lipoxin A4 robustly inhibit key events in acute inflammation, including neutrophil chemotaxis and transmigration across epithelial and endothelial cells, and down-regulate neutrophil diapedesis from post-capillary venules. Of more direct importance to the process of angiogenesis, 15 epi-lipoxin A4 and lipoxin A4 analogues inhibit cytokine release [58], and can modulate the local cytokine-chemokine axis [59]. Lipoxin A4 analogues do this by acting at the transcriptional level [60] and inhibiting VEGF-stimulated production of IL-6, tumor necrosis factor (TNF)-α, interferon (IFN)-α, interleukin (IL)-8, and inter-cellular adhesion molecule (ICAM)-1 expression [61]. However, although the aspirin-triggered 15 epi-lipoxin A4 has been shown to reduce polymorphonuclear cell infiltration and injury to mouse vasculature [62], we are not aware of any other publications documenting that this endogenously synthesized AA metabolite is an angiogenesis downregulator.

15 Deoxy-PGJ2

In agreement with our findings for the anti-angiogenesis activity of 15 deoxy-PGJ2, it was shown earlier that activation of PPAR-γ, but not PPAR-α or -β, by the specific ligand 15 deoxy-PGJ2 suppresses human umbilical vein endothelial cells' differentiation [63]. PPAR-γ activation also inhibits the proliferative response of HUVEC to exogenous growth factors in *in vitro* models and has *in vivo* down-regulating effects on angiogenesis in a rat cornea model [32]. Other studies showed that treatment of HUVEC with 15 deoxy-PGJ2 reduced mRNA levels of vascular endothelial cell growth factor receptors 1 (FLT1) and 2 (FLK/KDR) and urokinase plasminogen activator and increased plasminogen activator inhibitor-1 (PAI-1) mRNA, which might be responsible for its anti-angiogenesis activity [64]. Similarly, lipoxin A4 pretreatment of endothelial cells was shown to be associated with decrease of VEGF-stimulated VEGF receptor 2 (KDR/FLK-1) phosphorylation and downstream signaling events including activation of phospholipase C-gamma, ERK1/2, and Akt [64].

In contrast to single mechanism based anti-angiogenesis products such as anti-VEGF, the AA-derived angiogenesis inhibitors 15 epi-lipoxin A4 or 15 deoxy-PGJ2 demonstrated a broad spectrum anti-angiogenesis efficacy against various pro-angiogenesis stimulus including AA-derived or containing lipid products or known growth factors such as bFGF or VEGF. These data suggest angiogenesis modulating effects of AA-derived products may well be accessible drug targets for anti-angiogenesis as well as pro-angiogenesis therapies.

Conclusions

In summary, the data generated using an *in vivo* angiogenesis model demonstrate that interplay exists between potent lipid mediators derived from the long-chain, polyunsaturated, omega-6 arachidonic acid. Since these mediators are derived from metabolic pathways that intersect with the effects of medications, genetics, and other heterogeneous factors involved in human health and disease, these data suggest the need for research focused on shifting the metabolic milieu toward one that is most advantageous.

References

1. Smith SC Jr, Allen J, Blair SN, Bonow RO, Brass LM, Fonarow GC, Grundy SM, Hiratzka L, Jones D, Krumholz HM, Mosca L, Pasternak RC, Pearson T, Pfeffer MA, Taubert KA (2006) AHA/ACC guidelines for secondary prevention for patients with coronary and other athero-sclerotic vascular disease: 2006 update: endorsed by the National Heart, Lung, and Blood Institute. Circulation 113(19):2363–2372
2. Yusuf S, Reddy S, Ounpuu S, Anand S (2001) Global burden of cardiovascular diseases: Part I: general considerations, the epidemiologic transition, risk factors, and impact of urbanization. Circulation 104(22):2746–2753
3. Grundy SM, Cleeman JI, Merz CN, Brewer HB Jr, Clark LT, Hunninghake DB, Pasternak RC, Smith SC Jr, Stone NJ (2004) Implications of recent clinical trials for the national cholesterol education program adult treatment panel III guidelines. Circulation 110(2):227–239

4. Zheng ZJ, Croft JB, Giles WH, Mensah GA (2001) Sudden cardiac death in the United States, 1989–1998. Circulation 104(18):2158–2163
5. Albert CM, Chae CU, Grodstein F, Rose LM, Rexrode KM, Ruskin JN, Stampfer MJ, Manson JE (2003) Prospective study of sudden cardiac death among women in the United States. Circulation 107(16):2096–2101
6. Anderson RN (2001) Deaths: leading causes for 1999. Natl Vital Stat Rep 49(11):1–87
7. Fuster V, Moreno PR, Fayad ZA, Corti R, Badimon JJ (2005) Atherothrombosis and high-risk plaque: Part I: evolving concepts. J Am Coll Cardiol 46(6):937–954
8. Jemal A, Ward E, Hao Y, Thun M (2005) Trends in the leading causes of death in the United States, 1970–2002. JAMA 294(10):1255–1259
9. Jemal A, Tiwari RC, Murray T, Ghafoor A, Samuels A, Ward E, Feuer EJ, Thun MJ (2004) Cancer statistics. CA Cancer J Clin 54(1):8–29
10. Gralow J, Ozols RF, Bajorin DF, Cheson BD, Sandler HM, Winer EP, Bonner J, Demetri GD, Curran W Jr, Ganz PA, Kramer BS, Kris MG, Markman M, Mayer RJ, Raghavan D, Ramsey S, Reaman GH, Sawaya R, Schuchter LM, Sweetenham JW, Vahdat LT, Davidson NE, Schilsky RL, Lichter AS (2008) Clinical cancer advances 2007: major research advances in cancer treatment, prevention, and screening–a report from the American Society of Clinical Oncology. J Clin Oncol 26(2):313–325
11. Ishii I, Fukushima N, Ye X, Chun J (2004) Lysophospholipid receptors: signaling and biology. Annu Rev Biochem 73:321–354
12. Khurana R, Simons M, Martin JF, Zachary IC (2005) Role of angiogenesis in cardiovascular disease: a critical appraisal. Circulation 112(12):1813–1824
13. O'Brien ER, Garvin MR, Dev R, Stewart DK, Hinohara T, Simpson JB, Schwartz SM (1994) Angiogenesis in human coronary atherosclerotic plaques. Am J Pathol 145(4):883–894
14. Sueishi K, Yonemitsu Y, Nakagawa K, Kaneda Y, Kumamoto M, Nakashima Y (1997) Atherosclerosis and angiogenesis. Its pathophysiological significance in humans as well as in an animal model induced by the gene transfer of vascular endothelial growth factor. Ann N Y Acad Sci 811(311–322):322–314
15. Simons M, Ware JA (2003) Therapeutic angiogenesis in cardiovascular disease. Nat Rev Drug Discov 2(11):863–871
16. Celletti FL, Waugh JM, Amabile PG, Brendolan A, Hilfiker PR, Dake MD (2001) Vascular endothelial growth factor enhances atherosclerotic plaque progression. Nat Med 7(4):425–429
17. Moulton KS, Heller E, Konerding MA, Flynn E, Palinski W, Folkman J (1999) Angiogenesis inhibitors endostatin or TNP-470 reduce intimal neovascularization and plaque growth in apolipoprotein E-deficient mice. Circulation 99(13):1726–1732
18. Cuthbertson WF (1976) Essential fatty acid requirements in infancy. Am J Clin Nutr 29(5):559–568
19. Tapiero H, Ba GN, Couvreur P, Tew KD (2002) Polyunsaturated fatty acids (PUFA) and eicosanoids in human health and pathologies. Biomed Pharmacother 56(5):215–222
20. Iniguez MA, Cacheiro-Llaguno C, Cuesta N, Diaz-Munoz MD, Fresno M (2008) Prostanoid function and cardiovascular disease. Arch Physiol Biochem 114(3):201–209
21. Calder PC (2006) N-3 polyunsaturated fatty acids, inflammation, and inflammatory diseases. Am J Clin Nutr 83(6 Suppl):1505S–1519S
22. Gauthier KM, Yang W, Gross GJ, Campbell WB (2007 Dec) Roles of epoxyeicosatrienoic acids in vascular regulation and cardiac preconditioning. J Cardiovasc Pharmacol 50(6):601–608
23. Wray J, Bishop-Bailey D (2008) Epoxygenases and peroxisome proliferator-activated receptors in mammalian vascular biology. Exp Physiol 93(1):148–154
24. Levick SP, Loch DC, Taylor SM, Janicki JS (2007) Arachidonic acid metabolism as a potential mediator of cardiac fibrosis associated with inflammation. J Immunol 178(2):641–646
25. Aoki J (2004) Mechanisms of lysophosphatidic acid production. Semin Cell Dev Biol 15(5):477–489

26. Serhan CN, Chiang N, Van Dyke TE (2008) Resolving inflammation: dual anti-inflammatory and pro-resolution lipid mediators. Nat Rev Immunol 8(5):349–361
27. Hjelte LE, Nilsson A (2005) Arachidonic acid and ischemic heart disease. J Nutr 135(9):2271–2273
28. Glew RH, Okolie H, Huang YS, Chuang LT, Suberu O, Crossey M, VanderJagt DJ (2004) Abnormalities in the fatty-acid composition of the serum phospholipids of stroke patients. J Natl Med Assoc 96(6):826–832
29. Pozzi A, Macias-Perez I, Abair T, Wei S, Su Y, Zent R, Falck JR, Capdevila JH (2005) Characterization of 5,6- and 8,9-epoxyeicosatrienoic acids (5,6- and 8,9-EET) as potent in vivo angiogenic lipids. J Biol Chem 280(29):27138–27146
30. Mills GB, Moolenaar WH (2003) The emerging role of lysophosphatidic acid in cancer. Nat Rev Cancer 3(8):582–591
31. Cezar-de-Mello PF, Vieira AM, Nascimento-Silva V, Villela CG, Barja-Fidalgo C, Fierro IM (2008) ATL-1, an analogue of aspirin-triggered lipoxin A4, is a potent inhibitor of several steps in angiogenesis induced by vascular endothelial growth factor. Br J Pharmacol 153(5):956–965
32. Xin X, Yang S, Kowalski J, Gerritsen ME (1999) Peroxisome proliferator-activated receptor gamma ligands are potent inhibitors of angiogenesis in vitro and in vivo. J Biol Chem 274(13):9116–9121
33. Fleming I (2007) Epoxyeicosatrienoic acids, cell signaling and angiogenesis. Prostaglandins Other Lipid Mediat 82(1–4):60–67
34. Spector AA (2009) Arachidonic acid cytochrome P450 epoxygenase pathway. J Lipid Res 50(Suppl):S52–56
35. Pozzi A, Popescu V, Yang S, Mei S, Shi M, Puolitaival SM, Caprioli RM, Capdevila JH (2010) The anti-tumorigenic properties of peroxisomal proliferator-activated receptor alpha are arachidonic acid epoxygenase-mediated. J Biol Chem 285(17):12840–12850
36. Webler AC, Michaelis UR, Popp R, Barbosa-Sicard E, Murugan A, Falck JR, Fisslthaler B, Fleming I (2008) Epoxyeicosatrienoic acids are part of the VEGF-activated signaling cascade leading to angiogenesis. Am J Physiol Cell Physiol 295(5):C1292–1301
37. Michaelis UR, Falck JR, Schmidt R, Busse R, Fleming I (2005) Cytochrome P4502C9-derived epoxyeicosatrienoic acids induce the expression of cyclooxygenase-2 in endothelial cells. Arterioscler Thromb Vasc Biol 25(2):321–326
38. Wu WT, Chen CN, Lin CI, Chen JH, Lee H (2005) Lysophospholipids enhance matrix metalloproteinase-2 expression in human endothelial cells. Endocrinology 146(8):3387–3400
39. Gerrard JM, Robinson P (1989) Identification of the molecular species of lysophosphatidic acid produced when platelets are stimulated by thrombin. Biochim Biophys Acta 1001(3):282–285
40. Gaits F, Fourcade O, Le Balle F, Gueguen G, Gaige B, Gassama-Diagne A, Fauvel J, Salles JP, Mauco G, Simon MF, Chap H (1997) Lysophosphatidic acid as a phospholipid mediator: pathways of synthesis. FEBS Lett 410(1):54–58
41. Lee H, Liao JJ, Graeler M, Huang MC, Goetzl EJ (2002) Lysophospholipid regulation of mononuclear phagocytes. Biochim Biophys Acta 1582(1–3):175–177
42. Shen Z, Belinson J, Morton RE, Xu Y (1998) Phorbol 12-myristate 13-acetate stimulates lysophosphatidic acid secretion from ovarian and cervical cancer cells but not from breast or leukemia cells. Gynecol Oncol 71(3):364–368
43. Zhang C, Baker DL, Yasuda S, Makarova N, Balazs L, Johnson LR, Marathe GK, McIntyre TM, Xu Y, Prestwich GD, Byun HS, Bittman R, Tigyi G (2004) Lysophosphatidic acid induces neointima formation through PPARgamma activation. J Exp Med 199(6):763–774
44. Zhao Y, Natarajan V (2009) Lysophosphatidic acid signaling in airway epithelium: role in airway inflammation and remodeling. Cell Signal 21(3):367–377
45. Prestwich GD, Gajewiak J, Zhang H, Xu X, Yang G, Serban M (2008) Phosphatase-resistant analogues of lysophosphatidic acid: agonists promote healing, antagonists and autotaxin inhibitors treat cancer. Biochim Biophys Acta 1781(9):588–594

46. Bektas M, Payne SG, Liu H, Goparaju S, Milstien S, Spiegel S (2005) A novel acylglycerol kinase that produces lysophosphatidic acid modulates cross talk with EGFR in prostate cancer cells. J Cell Biol 169(5):801–811
47. Spiegel S, Merrill AH Jr (1996) Sphingolipid metabolism and cell growth regulation. FASEB J 10(12):1388–1397
48. Panchatcharam M, Miriyala S, Yang F, Rojas M, End C, Vallant C, Dong A, Lynch K, Chun J, Morris AJ, Smyth SS (2008) Lysophosphatidic acid receptors 1 and 2 play roles in regulation of vascular injury responses but not blood pressure. Circ Res 103(6):662–670
49. Moolenaar WH (1999) Bioactive lysophospholipids and their G protein-coupled receptors. Exp Cell Res 253(1):230–238
50. An S, Bleu T, Zheng Y, Goetzl EJ (1998) Recombinant human G protein-coupled lysophosphatidic acid receptors mediate intracellular calcium mobilization. Mol Pharmacol 54(5):881–888
51. Tigyi G, Fischer DJ, Sebok A, Marshall F, Dyer DL, Miledi R (1996) Lysophosphatidic acid-induced neurite retraction in PC12 cells: neurite-protective effects of cyclic amp signaling. J Neurochem 66(2):549–558
52. Stahle M, Veit C, Bachfischer U, Schierling K, Skripczynski B, Hall A, Gierschik P, Giehl K (2003) Mechanisms in LPA-induced tumor cell migration: critical role of phosphorylated ERK. J Cell Sci 116(Pt 18):3835–3846
53. Seewald S, Schmitz U, Seul C, Ko Y, Sachinidis A, Vetter H (1999) Lysophosphatidic acid stimulates protein kinase C isoforms alpha, beta, epsilon, and zeta in a pertussis toxin sensitive pathway in vascular smooth muscle cells. Am J Hypertens 12(5):532–537
54. Hayashi K, Takahashi M, Nishida W, Yoshida K, Ohkawa Y, Kitabatake A, Aoki J, Arai H, Sobue K (2001) Phenotypic modulation of vascular smooth muscle cells induced by unsaturated lysophosphatidic acids. Circ Res 89(3):251–258
55. Yoshida K, Nishida W, Hayashi K, Ohkawa Y, Ogawa A, Aoki J, Arai H, Sobue K (2003) Vascular remodeling induced by naturally occurring unsaturated lysophosphatidic acid in vivo. Circulation 108(14):1746–1752
56. Fierro IM, Kutok JL, Serhan CN (2002) Novel lipid mediator regulators of endothelial cell proliferation and migration: aspirin-triggered-15R-lipoxin A(4) and lipoxin A(4). J Pharmacol Exp Ther 300(2):385–392
57. Cezar-de-Mello PF, Nascimento-Silva V, Villela CG, Fierro IM (2006) Aspirin-triggered lipoxin A4 inhibition of VEGF-induced endothelial cell migration involves actin polymerization and focal adhesion assembly. Oncogene 25(1):122–129
58. Pouliot M, Serhan CN (1999) Lipoxin A4 and aspirin-triggered 15-epi-LXA4 inhibit tumor necrosis factor-alpha-initiated neutrophil responses and trafficking: novel regulators of a cytokine-chemokine axis relevant to periodontal diseases. J Periodontal Res 34(7):370–373
59. Hachicha M, Pouliot M, Petasis NA, Serhan CN (1999) Lipoxin (LX)A4 and aspirin-triggered 15-epi-LXA4 inhibit tumor necrosis factor 1alpha-initiated neutrophil responses and trafficking: regulators of a cytokine-chemokine axis. J Exp Med 189(12):1923–1930
60. Gewirtz AT, McCormick B, Neish AS, Petasis NA, Gronert K, Serhan CN, Madara JL (1998) Pathogen-induced chemokine secretion from model intestinal epithelium is inhibited by lipoxin A4 analogs. J Clin Invest 101(9):1860–1869
61. Teutsch SM, Cohen JT (2005) Health trade-offs from policies to alter fish consumption. Am J Prev Med 29(4):324
62. Takano T, Clish CB, Gronert K, Petasis N, Serhan CN (1998) Neutrophil-mediated changes in vascular permeability are inhibited by topical application of aspirin-triggered 15-epi-lipoxin A4 and novel lipoxin B4 stable analogues. J Clin Invest 101(4):819–826
63. Imaizumi T, Matsumiya T, Tamo W, Shibata T, Fujimoto K, Kumagai M, Yoshida H, Cui XF, Tanji K, Hatakeyama M, Wakabayashi K, Satoh K (2002) 15-Deoxy-D12,14-prostaglandin J2 inhibits CX3CL1/fractalkine expression in human endothelial cells. Immunol Cell Biol 80(6):531–536
64. Kim EH, Na HK, Surh YJ (2006) Upregulation of VEGF by 15-deoxy-delta12,14-prostaglandin J2 via heme oxygenase-1 and ERK1/2 signaling in MCF-7 cells. Ann N Y Acad Sci 1090:375–384

Chapter 4
Pro-angiogenic Activity of Thyroid Hormone Analogues: Mechanisms, Physiology and Clinical Prospects

Paul J. Davis, Faith B. Davis, Hung-Yun Lin, Mary K. Luidens, and Shaker A. Mousa

Abstract Thyroid hormone and analogues have demonstrated potent pro-angiogenic activity in various *in vitro* and *in vivo* model systems. These pro-angiogenic effects of thyroid hormone are initiated nongenomically at a receptor on integrin $\alpha v\beta 3$ on receptor blood vessel endothelial cells, but may culminate in genomic responses. This receptor on the cell surface may direct gene transcription and regulate activity of plasma membrane vascular growth factor receptors. Promotion of hormonal angiogenic activity can be blocked by tetrac and nanoparticulate tetrac that inhibit binding of T_4 and T_3 to $\alpha v\beta 3$, and also block activities of endogenous vascular growth factors. Endogenous or prescribed thyroid hormone appears to support tumor-relevant angiogenesis. It may also support pathologic skin neovascularization. In the absence of malignancy, however, the angiogenic activity of the hormone has exhibited experimental utility in wound-healing and tissue ischemia. Clinical applications of the hormone may be optimized by a nanoparticulate to which the hormone is covalently bound. Nanoparticulate thyroid hormone does not gain access to the intracellular space and acts exclusively at the hormone receptor on integrin $\alpha v\beta 3$. Restriction of the hormone to the extracellular space avoids

P.J. Davis (✉)
Department of Medicine, Albany Medical Center, Albany, NY, USA

The Pharmaceutical Research Institute at Albany College of Pharmacy
and Health Sciences, Rensselaer, NY, USA
e-mail: pdavis.ordwayst@gmail.com

F.B. Davis • S.A. Mousa
The Pharmaceutical Research Institute at Albany College of Pharmacy
and Health Sciences, Rensselaer NY, USA

H.-Y. Lin
Institute of Cancer Biology and Drug Discovery, Taipei Medical University, Taipei, Taiwan

M.K. Luidens
Department of Medicine, Albany Medical College, Albany, NY, USA

S.A. Mousa and P.J. Davis (eds.), *Angiogenesis Modulations in Health and Disease:*   29
Practical Applications of Pro- and Anti-angiogenesis Targets,
DOI 10.1007/978-94-007-6467-5_4, © Springer Science+Business Media Dordrecht 2013

unwanted hormonal effects on cellular respiration, on protein synthesis and on protein turnover and specific organ effects, such as cardiac tachyarrhythmias. Needed are more extensive preclinical data on the actions of thyroid hormone, T_4 and T_3 on, respectively, wound-healing and revascularization of limb or cardiac ischemic tissues.

History of Thyroid Hormone-induced Angiogenesis

Actions of thyroid hormone (L-thyroxine, T_4 and 3, 5, 3′-triiodo-L-thyronine, T_3) on blood vessels have been described clinically for many decades. The hormone may widen pulse pressure by reducing vascular resistance and increasing myocardial contractility. Cutaneous vasodilatation is a component of hyperthyroidism. Acute administration of the hormone to human subjects can affect peripheral vascular resistance in minutes [1]. That the hormone is pro-angiogenic was convincingly first shown by W.M. Chilian and co-workers (1985) [2] who were examining thyroid hormone-induced myocardial hypertrophy in the rat. The investigators found capillary mass/mm^2 increased in the L-thyroxine (T_4)-treated animals and, in a subsequent study, showed that T_4 caused an increase in capillary numerical density before hypertrophy was manifest [3]. Several members of the same group have also reported that a thyroid hormone analogue, diiodothyropropionic acid (DITPA), induces increases in terminal arteriole number and arteriole length density, but without increases in cardiac mass [4]. Structures of T_4, DITPA and T_3 are shown in Fig. 4.1.

Studies carried out in the past decade in the chick chorioallantoic membrane (CAM) model system confirmed that T_4 and T_3 were pro-angiogenic [5, 6], as did additional experiments with human dermal microvascular endothelial cells (HDMECs) in the three-dimensional angiogenic Matrigel® assay [7]. These observations initiated a series of studies of additional iodothyronine analogues for their activity in the CAM assay and studies of the molecular basis for this nongenomic action of members of the thyroid hormone family. The studies included identification of the previously unrecognized cell surface receptor for thyroid hormone on a polyfunctional structural plasma membrane protein, integrin αvβ3. Antagonists of agonist thyroid hormone family members at the integrin have also been described. These studies are reviewed below.

Thyroid Hormone Receptor Site on Integrin αvβ3

The classical or genomic mechanism of thyroid hormone action requires cell nucleus uptake of T_3 and binding of the iodothyronine by one of several hormone-specific nuclear transcription factors. These factors—nuclear thyroid hormone receptors (TRs), TRβ1 and TRα1—bind to thyroid hormone response elements (TREs) of hormone-responsive genes to initiate transcription [8]. TR-directed

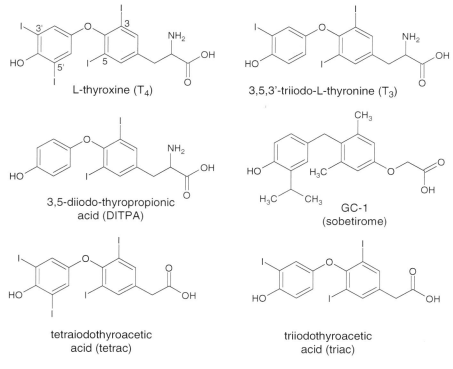

Fig. 4.1 Structures of thyroid hormone (L-thyroxine, T$_4$; 3, 5, 3'-triiodo-L-thyronine, T$_3$) and hormone analogues discussed in this chapter

transcription is modulated by co-activators, such as p300, or co-repressors, such as NCoR or SMRT [8]. Expression of 100 or more genes is regulated by this genomic mechanism that depends upon the T$_3$-TR complex. It is important to emphasize that tissue and blood levels of T$_3$ are relatively stable, and thus the contribution of the hormone to gene expression is not an off-on or cyclical phenomenon, but rather as a setpoint regulator. In this mechanism, circulating T$_4$ is a pro-hormone, yielding T$_3$ via tissue deiodinase activity [8].

In vitro studies in the 1980s supported the existence of a second and nongenomic mechanism of thyroid hormone action. Effects of the hormone that necessarily were independent of TR were those reported on ion transporters in isolated membranes, e.g., on Ca^{2+}-ATPase (calcium pump) [9, 10] activity in human red blood cell membranes or in animal cardiac sarcoplasmic reticulum [11]. Subsequently, effects of T$_3$ on Na/H exchange [12] and on Na, K-ATPase (sodium pump) activity [13] were reported that also did not depend upon gene transcription. Other effects of the hormone were described that occurred so rapidly—minutes to an hour—after exposure of the test biologic preparation to thyroid hormone *in vitro* that the processes of gene transcription and protein synthesis could not be implicated. Among such iodothyronine effects were actions on glucose transport [14], intracellular shuttling of proteins [15], actin polymerization and the inward rectifying myocardial potassium channel [16, 17]. Actin polymerization was shown to be unresponsive to T_3, but promoted by T_4 [18, 19]. Thus, in nongenomic mechanisms of thyroid hormone action, T_4 can function as a hormone rather than as a pro-hormone [8].

The discovery of a receptor for thyroid hormone analogues on the cell surface that could be linked to some of the processes listed above occurred in 2005 [20]. Recognition and definition of the site was the result of the re-investigation of the pro-angiogenic activity of thyroid hormone [5] in biologic assay systems that depended upon the presence of integrin $\alpha v\beta 3$. Integrins are structural proteins of the plasma membrane. Computer-based crystallographic modeling of $\alpha v\beta 3$ allowed the localization of the receptor to a site in the extracellular domain of the integrin and to which both the αv and $\beta 3$ monomers contributed [21, 22]. Prior to the recognition on $\alpha v\beta 3$ of the small molecule receptor for iodothyronines, the integrin was seen exclusively to have large molecule, i.e., protein, ligands from extracellular matrix [23]. Examples of such proteins are vitronectin, fibronectin and von Willebrand factor. While there are more than 20 integrins, we have found radiolabeled thyroid hormone-binding to occur to only one cell membrane protein, the $\alpha v\beta 3$ integrin (J.J. Bergh, P.J. Davis: unpublished observations).

The receptor site we identified was specific for, and had high affinities for, thyroid hormone analogues [8, 24]. The site was also able to distinguish among T_4, T_3 and the deaminated T_4 analogue, tetraiodothyroacetic acid (tetrac) (Fig. 4.1); that is, certain of the cellular consequences of the binding of such analogues were distinct. Blocking of binding of T_4 and T_3 to the receptor site inhibited the pro-angiogenic activity of these forms of thyroid hormone. The site was subsequently shown to be capable of, and differentially so, of activating mitogen-activated protein kinase (MAPK; extracellular regulated kinase 1/2, ERK1/2) or phosphatidylinositol 3-kinase (PI3K) to achieve discrete downstream intracellular or tissue consequences, such as angiogenesis, tumor cell proliferation, transcription of specific genes and trafficking of intracellular proteins [24].

The hormone-binding site on the integrin has two binding domains, one (S1) that recognizes T_3 exclusively and a second (S2) that binds both T_4 and T_3. The existence of the two domains has been confirmed by hormone analogue-binding kinetics [25] and crystallographic modeling [22]. The receptor site engages in crosstalk with vascular growth factor receptors that are juxtaposed with the integrin and is also capable of stimulating the phosphorylation of estrogen receptor-α (ERα) and inducing proliferation of ERα-dependent breast cancer and lung cancer cells. We now know that the cell surface receptor on $\alpha v\beta 3$ also modulates expression of differentially regulated genes important to cancer cell biology [26]. Such findings—those related to nuclear ERα and to expression of specific genes—blur the distinction between genomic and nongenomic actions of thyroid hormone. That is, actions of T_4 and T_3 and other hormone analogues can be initiated nongenomically at the cell surface and conclude with nucleoprotein phosphorylation and gene expression, some of which involves activation of nuclear receptor proteins for estrogen and for thyroid hormone.

When it gains access to the cell interior, agonist thyroid hormone T_3 has effects on cell respiration, protein synthesis and degradation, and on expression of a large number of genes. In specialized tissues such as the myocardium, clinically administered T_3 may induce cardiac arrhythmias. For angiogenic purposes, thyroid hormone has been reformulated as a nanoparticle that acts exclusively at the $\alpha v \beta 3$ receptor for the hormone. We have covalently attached agonist thyroid hormone by its distal ring hydroxyl to biodegradable poly(lactic-co-glycolic acid) (PLGA). The resulting molecule is too large to gain access to the cell interior. Potency of the hormone at the plasma membrane receptor is preserved or even enhanced in the nanoparticle.

We have described other small molecule receptor sites on integrin $\alpha v \beta 3$, for example, a site that binds resveratrol [27] and one for dihydrotestosterone [28]. These sites qualify as receptors because of ligand specificity and well-defined downstream consequences of the binding of their ligands. These small molecule receptors for thyroid hormone, for the stilbene (resveratrol), and for androgen, apparently are not linked to one another. That is, there is not mutual competition among the specific ligands for the group of small molecule receptors. The receptors nonetheless are geographically close to the Arg-Gly-Asp (RGD) recognition site on the integrin that is important to extracellular matrix protein-binding by $\alpha v \beta 3$ [21], and RGD peptides can sometimes affect the small molecule receptors. RGD peptides differentially block activities of the S1 and S2 domains of the thyroid hormone receptor [25].

The pro-angiogenic activity of thyroid hormone analogues that has been well-described [29, 30] is a function of the plasma membrane receptor and confirms the generous expression of integrin $\alpha v \beta 3$ on rapidly dividing blood vessel cells [31]. Interestingly, T_4 is pro-angiogenic without prior conversion to T_3. The integrin also can serve as a radiographic target that enables the localization of exuberant blood vessel formation that is supporting tumors [32, 33] and thus identification of small cancers.

Additional details of the molecular basis for angiogenesis promoted by thyroid hormone are provided below in Section "Molecular Basis for Thyroid Hormone-Induced Angiogenesis".

Aside from blood vessel cells, on what other cells is the integrin—and thus the thyroid hormone receptor—expressed in increased amounts? Integrin $\alpha v \beta 3$ is expressed generously on tumor cells and on osteoclasts. The integrin is found on nonmalignant cells in addition to osteoclasts and is functional, e.g., on platelets [34]. Its presence on osteoclasts would suggest a molecular basis for bone resorption that is seen in the setting of excess circulating thyroid hormone in the intact organism. T_4 induces osteoporosis in the rabbit and this effect of the hormone was shown to be blocked by RGD peptide administration [35] long before the receptor for T_4 and T_3 on $\alpha v \beta 3$ was described. The expression of integrin $\alpha v \beta 3$ on cancer cells has several implications that are discussed in Section "Tumor-relevant Angiogenesis and Thyroid Hormone Formulations; Thyroid Hormone and Tumor Cell Proliferation".

The integrin receptor for thyroid hormone is a potential vehicle in patients without cancer for management of disease processes that require neovascularization. These processes include narrowing of blood vessels in the extremities and in the coronary arteries. In a rabbit model of hind limb ischemia, infusion of T_4 was found by angiographic monitoring to stimulate new blood vessel buds and, on histologic examination, to increase the number of blood vessels/muscle fiber [36] (see Section "Vascular Disease in Limbs and Heart: Applications of Thyroid Hormone" below). Coating of stents with releasable thyroid hormone may be considered experimentally in coronary and limb arteries.

Topical application of T_3 has been shown experimentally to expedite wound-healing (see Section "Wound-healing and Thyroid Hormone" below).

In summary, the definition of the cell surface receptor for thyroid hormone in the extracellular domain of integrin $\alpha v \beta 3$ offers new opportunities for regulation of angiogenesis. The molecular basis for the pro-angiogenic action of the hormone is largely understood and the mechanisms are reviewed in the next section ("Molecular Basis for Thyroid Hormone-induced Angiogenesis"). Angiogenic thyroid hormone may be reformulated as a nanoparticle that acts exclusively at the integrin receptor for the hormone and does not gain entry into cells.

Molecular Basis for Thyroid Hormone-induced Angiogenesis

Agonist forms of thyroid hormone, such as T_4 and diiodothyropropionic acid (DITPA), have been shown to increase *bFGF* expression [4, 5], as well as *VEGF* [4] in cells and tissues. T_4 also stimulates the release in target cells of bFGF that acts in an autocrine fashion in the chick chorioallantoic membrane (CAM) model of angiogenesis to induce neovascularization. We also know that the thyroid hormone receptor site on $\alpha v \beta 3$ is capable of modulating angiogenic activities of exogenous VEGF and bFGF added to the CAM [37], as well as of platelet-derived growth factor (PDGF) and epidermal growth factor (EGF) (S.A. Mousa: unpublished). We have proposed that such effects are evidence for the existence of crosstalk between the integrin and adjacent receptors for VEGF (VEGFR) and other growth factors. Indeed, the crosstalk may be between the extracellular domains of the growth factor receptor and $\alpha v \beta 3$ in the case of PDGF [38] or the intracellular, sub-plasma membrane biochemical transactions between the intracellular domains of the two proteins in the case of VEGF [39]. Pharmacologic inhibition of MAPK disrupts modulation by the integrin of thyroid hormone analogue-modulated interactions of vascular growth factors with their specific receptors that are adjacent to $\alpha v \beta 3$.

What are the essential structural features of the thyroid hormone analogues that are pro-angiogenic at $\alpha v \beta 3$? In the CAM assay, T_4 and T_3 are angiogenic, as are DITPA [40] and a novel hormone analogue, GC-1 (sobetirome) [41], that contains no halogens. In GC-1, inner ring iodines are replaced by methyl groups, and the outer ring includes an isopropyl substitution for the 5' iodine; the alanine side chain is replaced with an oxyacetic chain, and the inter-ring ether linkage is replaced by a methylene bridge (Fig. 4.1). In the T_4 analogue, tetraiodothyroacetic acid (tetrac) (Fig. 4.1), the alanine side chain is replaced by acetic acid; in triiodothyroacetic acid (triac), derived from T_3, the side chain is also acetic acid. Tetrac and triac are anti-angiogenic by several mechanisms that are reviewed in Chap. 10, including inhibition of binding of agonist thyroid hormone analogues to the $\alpha v \beta 3$ thyroid hormone receptor site.

The structural features of thyroid hormone that are required for angiogenic activity are summarized in Table 4.1. It should be noted that the outer ring hydroxyl is shared by most of the hormone analogues depicted in Fig. 4.1, but the hydroxyl is not required for angiogenic activity. This has been shown when nanoparticulate formulations of T_4 were generated and were active in the CAM. The covalent bonding of T_4 to the nanoparticle was via an ether linkage at the hydroxyl site. The structure of the side chain on the inner ring is critical. In naturally occurring thyroid hormone analogues, the side chain is alanine, but in GC-1 the side chain is an acetate ether bonded to the ring. If the side chain length is shortened to only the acetic acid residue, analogue molecules are obtained which, as noted above, *inhibit* T_4- and T_3-binding to $\alpha v \beta 3$.

The relatively short list of shared features identified in Table 4.1 indicate that opportunities exist for design of additional iodothyronine-like molecules with angiogenic activity.

Nanoparticulate Thyroid Hormone Analogues

The initial proof that the pro-angiogenic activity of T_4 was initiated at the cell surface depended upon use of the hormone covalently bound to a large polysaccharide moiety (agarose) that prevented cellular uptake of T_4 [5]. Subsequently, S.A. Mousa and co-workers covalently linked thyroid hormone analogues to poly(lactic-co-glycolic acid) (PLGA), a biodegradable polymer [42, 43]. PLGA has been used clinically to change the pharmacokinetics of drugs noncovalently imbedded in the polymer and progressively released by the circulating or localized polymer. The approximately 200 nm thyroid hormone analogue-PLGA molecule does not gain access to the intracellular space [43, 44], but is biologically active at the hormone receptor on integrin $\alpha v \beta 3$. The assembly of the PLGA-hormone formulation is via a linker that is ether-bonded to the outer ring hydroxyl and amide-bonded to the PLGA, with the bond imbedded in the polymer and not readily accessible to circulating peptidases.

The advantages of the nanoparticulate formulation are several. First, and as noted above, the activity of the hormone ligand is restricted to the outer surface of the plasma membrane and, functionally, to integrin $\alpha v \beta 3$. There are no intracellular effects of the formulation(s). Second, the polymer may bear a payload of a second molecule unrelated to thyroid hormone that can be delivered to $\alpha v \beta 3$-expressing cells. Third, the polymer may lengthen the biological half-life of the hormone analogue to that of the time required for the degradation of the polymer to glycolic acid and lactic acid.

Table 4.1 Common structural features of thyroid hormone analogues that are angiogenic in the chick chorioallantoic membrane (CAM) assay

2 Aromatic (benzene) rings
Bridge between the rings, e.g., ether bond or methylene bridge
3-Carbon side chain with terminal carboxyl (see narrative)
Iodine-sized mass substituent at 3 and 5 (inner ring) positions

Finally, the polymer may change the presentation of the bound hormone analogue to the receptor site in a favorable manner. We have found, for example, that tetrac covalently bound to PLGA is more potent than unmodified tetrac and has discrete, additional biological effects, for example, on gene transcription [26].

Tumor-relevant Angiogenesis and Thyroid Hormone Formulations: Thyroid Hormone and Tumor Cell Proliferation

The pro-angiogenic activity of thyroid hormone has been demonstrated in a number of models and this activity presumptively supports tumor-related angiogenesis in the intact host. One experimental approach to the evaluation of the significance of this hormonal effect has to been to use tetrac in the setting of human cancer xenografts in the nude mouse. Tetrac and its nanoparticulate formulation are anti-angiogenic by blocking the actions of T_4 and T_3 at the hormone receptor on integrin $\alpha v \beta 3$ and, as pointed out above, by inhibiting the actions of several vascular growth factors released by tumor cells. In xenograft studies, we have found that tetrac formulations cause a more than 50 % decrease in cancer-supporting vascular volume within 3 days of initiation of treatment [44–47]. This anti-angiogenic activity of tetrac formulations may emerge as desirable in the clinical setting of cancer. Single target anti-angiogenic agents, e.g., the monoclonal antibody to VEGF (bevacizumab [Avastin®]), have shown clinical usefulness in combination with traditional agents in management of certain solid tumors [48, 49]. Tyrosine kinase inhibitors (TKIs) are anti-cancer agents that may also have anti-angiogenic properties [50]. An advantage of tetrac formulations over agents such as bevacizumab is that the tetrac and nanoparticulate tetrac affect the actions of multiple vascular growth factors.

In addition to supporting neovascularization about tumors, T_4 at physiologic concentrations is a proliferative factor for a variety of tumor cells *in vitro* [51–53]. T_3 may affect tumor cell biology only when present in supraphysiologic concentrations [51]. Induction of hypothyroidism pharmacologically improves survival in patients with advanced glioblastoma [54], and human breast cancer has less aggressive behavior when it occurs in the setting of spontaneous hypothyroidism [55]. The usefulness of TKI drugs in patients with kidney cancer (renal cell carcinoma) appears to be limited to those subjects who develop hypothyroidism as a complication of TKI management [56–58]. Hyperthyroidism may increase the risk of esophageal cancer [59], and the risk of breast cancer appears to be increased in women with higher serum thyroid hormone (FT_4) levels [60]. These recent observations suggest that the integrin $\alpha v \beta 3$ receptor can mediate actions of circulating thyroid hormone on tumor behavior. Thus, we recommend that clinical treatment of spontaneous hypothyroidism in the presence of cancer should be carefully designed to limit symptoms of hypothyroidism, rather than to reduction of the serum thyrotropin (TSH) concentration to a specific level in the reference range. In addition to its effects on cancer cells, this approach further serves to limit the pro-angiogenic contribution of endogenous thyroid hormone to the nurture of established cancers.

Vascular Disease in Limbs and Heart: Applications of Thyroid Hormone

Conceptually, clinical management of pathological narrowing of blood vessels due to atherosclerosis should involve several steps. First, where feasible, sites of critical narrowing should be relieved. Second, local and regional angiogenesis should be fostered as a therapeutic and preventive measure. The angiogenesis should be detectable in regional muscle mass.

El-Eter and co-workers [36] have studied thyroid hormone-directed angiogenesis in an animal model of leg ischemia. Perfusion of the affected limb in the rabbit with T_4 resulted in vascular sprouting or buds, detected by limb angiography. Further, muscle histology in the T_4-exposed limb showed significantly increased numbers of capillaries/muscle fiber.

Tomanek et al. [61, 62] have studied experimental left ventricular infarction in rats treated with DITPA. Angiogenesis quantitated as capillary-arteriolar length density was found to be increased in DITPA-exposed animals in the border regions of myocardial infarcts, particularly, large infarctions. These pro-angiogenic changes limited infarct expansion and undesirable post-infarction ventricular remodeling that may lead to heart failure.

In a non-ischemic hypothyroid rat heart model, administration of T_3 has recently been shown to promote small arteriole muscularization and to effectively return myocardial arteriolar density to normal [63]. Effects are maximal within 72 h.

Thus, in experimental animal models, local administration of thyroid hormone (T_4, DITPA, T_3) can produce anatomically desirable angiogenesis in ischemic vessels in limbs and in the heart. The pro-aggregatory effect of T_4 on platelets [34], however, is undesirable in the ischemic, low blood flow setting.

Wound-healing and Thyroid Hormone

The contribution of thyroid hormone to wound-healing has been studied for more than 15 years. The reports available, however, have not systematically examined the neovascularization that supports healing. For example, Natori et al. in 1999 [64] described the healing in hypothyroid rats of sutured abdominal incisions. Healing was delayed in the hypothyroid animals, compared with controls. Decreased type IV collagen and hydroxyproline were found at the wound sites in hypothyroid rats during the proliferative phase of healing and were thought to contribute to healing delay.

Safer, Crawford and Holick in 2005 [65] applied T_3 topically to skin wounds of the backs of mice and reported significantly (P <0.001 vs. controls) greater wound closure in the hormone-treated animals. Closure was measured directly over time as the diameter of circular wounds. This improvement was attributed at least in part to increased local accumulation of keratin 6 in keratinocytes. The same authors had previously described a *keratin* gene response (K6a, K16, K17) in keratinocytes exposed *in vitro* to T_3 [66]. The authors proposed that topical T_3 may come to be viewed as an alternative measure in hypothyroid surgical patients for whom insufficient time exists preoperatively to repair the hypothyroid state. Safer and colleagues also suggested that topical T_3 be examined as an inexpensive measure to expedite healing in euthyroid subjects.

Subsequently, Tha Nassif et al. [67] studied a model of colonic anastomosis-healing in hypothyroid and control rats. In the hypothyroid animals, impaired healing was documented by decreased collagen density, increased local accumulation of immature (type III) collagen and decreased accumulation of mature (type 1) collagen. A functional, quantifiable consequence of this was a reduced value in the hypothyroid group for the bursting strength test. Slowed wound-healing and decreased local accumulation of collagen has also been described in tracheal wounds in hypothyroid animals [68].

The healing process after leg muscle ischemia-reperfusion injury displays a more complex response in hypothyroid animals, compared to rats with intact thyroid function. Ozawa and co-workers [69] compared muscle repair in the setting of (1) 4 h of ischemia, then reperfusion, and (2) surgical transection of soleus muscle in hypothyroid and control animals. In the ischemia-reperfusion model, there was muscle necrosis in the euthyroid rats and only modest muscle damage in the hypothyroid animals at the time the animals were sacrificed. In transected muscle, hypothyroidism impaired debris removal and slowed myotubule formation, so that it is presumed the normal process of repair in muscle damaged by ischemia was delayed in hypothyroidism.

We can conclude from this body of experimental evidence that direct application to wounds of thyroid hormone as T_3 expedites healing. A variety of studies in intact animals also suggest that normal, stable circulating levels of thyroid hormone support an organized process of healing in complex wounds, for example, those of the bowel or airway. The contributions of the hormone to the wound-healing process appear to involve fibroblasts—the source of collagen—and, in the skin, keratinocytes. One presumes that thyroid hormone-induced angiogenesis, discussed elsewhere in this chapter, locally supports the healing process, but quantitative information on this topic is needed.

Is There a Contribution of Thyroid Hormone to Brain Angiogenesis in the Embryo?

Little is known about the possible contributions of thyroid hormone to angiogenesis in the embryo or fetus. Brain blood vessel formation during brain development is a principal concern. Angiogenesis has been studied from birth to postnatal day 90 (P90) in rats rendered hypothyroid by propylthiouracil (PTU) started at birth [70]. Decreased angiogenesis (reduced complexity and density of microvessels) was present in brain by P21, but withdrawal of PTU at that time resulted in complete recovery of vascularization by P90.

In a similar study, Schlenker et al. [71] studied blood vessel density (BVD) in adult rat brains as a function of thyroid state. By 6 weeks post-thyroidectomy, rats had decreased forebrain BVD compared to euthyroid control animals. Rats treated from the time of thyroidectomy with T_4 or DITPA exhibited no BVD diminution. These studies securely document the pro-angiogenic impact of thyroid hormone analogues on brain blood vessels. Issues to be addressed experimentally in this area of research include specific examination of the contribution of thyroid hormone to angiogenesis during brain development *in utero* and, in the mature organism, whether thyroid hormone can limit neuronal injury in the setting of ischemia.

Hyper- and Hypothyroidism and Angiogenesis

Little systematically collected information is available regarding possible effects of the clinical states of hyperthyroidism and hypothyroidism on angiogenesis. Changes in wound-healing that may be observed in hypothyroidism have been discussed in Section "Wound-healing and Thyroid Hormone", and the loss of capillary density and arteriolar smooth muscle mass in the heart in experimental hypothyroidism were described in Section "Vascular Disease in Limbs and Heart: Applications of Thyroid Hormone". T_4 has been shown to increase rat ovarian follicle angiogenesis [72]. While increased intrathyroidal angiogenesis has been documented in Graves' hyperthyroidism [73, 74] and autoimmune thyroiditis [74] (Graves' > thyroiditis), it is not possible in these studies to dissociate possible effects of thyroid hormone from those of pituitary thyrotropin (TSH) in thyroiditis and of human thyroid-stimulating immunoglobulin in the setting of Graves' disease.

Is Diabetic Retinopathy Supported by Thyroid Hormone?

Diabetic retinopathy has not been studied when spontaneous hypo- or hyperthyroidism have supervened clinically in diabetic patients. Three recent reports have reported an association between subclinical hypothyroidism—asymptomatic patients with elevated serum thyrotropin (TSH) levels and normal circulating free T_4 and total T_3—and worsening proliferative retinopathy [75–77]. This is an interesting finding and is not consistent with the pro-angiogenic activity credibly ascribed to T_4, T_3 and other agonist thyroid hormone analogues. It is also not clear in these reports whether the patients were receiving supplemental T_4 as treatment to reduce TSH. On the other hand, TSH has been shown to stimulate VEGF production in thyroid cancer [78] and to induce angiogenesis and VEGF accumulation in the human normal thyroidal xenograft [79]. However, it is not known whether TSH can stimulate angiogenesis in nonthyroidal tissues.

We point out that tetrac, an antagonist of thyroid hormone action at the thyroid hormone receptor on $\alpha v \beta 3$, inhibits the action of thyroid hormone on angiogenesis (see Chap. 10) and has been shown to inhibit experimental proliferative retinopathy in the mouse [80]. This suggests that endogenous thyroid hormone may play a pathogenetic role in such retinopathy, but, independent of the presence or absence of thyroid hormone, tetrac also blocks the actions of VEGF and bFGF (see Chap. 10).

Thus, systematic analysis of diabetic populations in which hypothyroidism or hyperthyroidism has occurred is required to clarify any relationship that may occur between clinical thyroid dysfunction and retinopathy.

Neovascularization and Inflammation: Role of Thyroid Hormone?

The inflammatory process depends upon local accumulation of white blood cells, cytokines and neovascularization. In the case of certain skin inflammation states, such as acne rosacea, an inciting influence may not be defined. The downstream mechanisms involved, however, usually involve local release of traditional vascular growth factors, such as VEGF. Smith et al. [81] have described the expression of VEGF receptors (VEGFR-1, VEGFR-2) in skin involved with rosacea. The sources of the VEGF, itself, may be infiltrating inflammatory cells, for example, granulo- cytes [82] or cells of the skin layers (fibroblasts [83]). Neutrophils also express basic fibroblast growth factor (bFGF) [84], another pro-angiogenic agent, and this may be relevant to rosacea or to psoriasis [85].

Elsewhere in this chapter, we have cited evidence that thyroid hormone can potentiate the angiogenic activities of VEGF and bFGF [7] or may stimulate local expression and release of bFGF [5]. Such actions involve nongenomic mechanisms that we now know begin at the cell surface receptor for thyroid hormone on integrin $\alpha v\beta 3$ [20] and involve crosstalk with receptors on the plasma membrane for specific vascular growth factors. These observations suggest that circulating, usually stable, levels of thyroid hormone support the erythema of acne rosacea and other skin con- ditions marked by exuberant angiogenesis. We also know, as pointed out above, that the pro-angiogenic activity of platelet-derived growth factor (PDGF) is potentiated by iodothyronines (S.A. Mousa: unpublished). Actions of epidermal growth factor (EGF) may be enhanced by thyroid hormone [86], and this factor also stimulates angiogenesis [87]. These issues aside, there is no information in the literature on the effects of spontaneous hypothyroidism or hyperthyroidism on the activity of acne rosacea.

The expense and requirement for systemic administration of monoclonal VEGF antibody (Avastin®) or EGF antibody (Erbitux®, cetuximab) render impractical the use of such agents in skin inflammatory states. Tetraiodothyroacetic acid (tetrac) inhibits activities of T_4 and T_3 at their receptor on integrin $\alpha v\beta 3$ and has been shown to inhibit activities of the principal vascular growth factors [7]. It can be adminis- tered topically and thus may be a candidate drug for management of VEGF- and bFGF- dependent disorders of the skin. Tetrac and its nanoparticulate formulation are discussed in Chap. 10.

Medically important arteriovenous malformations (AVMs) may also express VEGF locally [88], notably in brain AVMs [89]. As in the case of the skin, it is pos- sible that stable circulating levels of thyroid hormone may support or enhance the local activity of VEGF in the disease process. No information currently exists in the literature to link clinical thyroid gland disorders—hyper- and hypothyroidism—to the course of disease in AVMs.

Non-thyroidal Illness Syndrome and Angiogenesis

The non-thyroidal illness (NTI) syndrome or 'euthyroid sick syndrome' complicates clinically important acute and chronic disease states. A significant minority of hospital inpatients exhibit low circulating levels of T_3 in response to their non-thyroidal diseases and that reflects, in multiple organs, the downregulated deiodination of blood and tissue T_4 to generate T_3 [90]. Circulating total and free T_4 may be normal, or the free T_4 may be transiently elevated. A review of the tissue deiodinase pathophysiology involved and the conjectural utility of the syndrome—for example, possible slowing of protein turnover in the face of physical stress—is beyond the scope of this chapter. However, the existence and high incidence of the NTI syndrome raises the issue of whether angiogenesis is adversely affected by the low systemic levels of T_3 in the NTI setting. Slowed wound-healing (Section "Wound-healing and Thyroid Hormone") or undesirable myocardial remodeling after infarction [91] might be consequences of low tissue T_3 levels. To date, these possibilities have not been examined experimentally.

Conclusions

Examined *in vitro* or *in vivo*, thyroid hormone has substantial pro-angiogenic activity. Thyroid hormone family members with such activity are T_4, T_3, DITPA and GC-1. The nongenomic molecular basis for this action of the hormone has been elucidated and begins with hormone-binding by a plasma membrane receptor for hormone analogues on integrin $\alpha v \beta 3$ and requires activation of MAPK (ERK1/2) and, downstream of the receptor and kinase, vascular growth factor production. Promotion of hormonal angiogenic activity can be blocked by tetrac and nanoparticulate tetrac, agents that inhibit binding of T_4 and T_3 to $\alpha v \beta 3$. Tetrac is the deaminated derivative of T_4. Triac, the deaminated metabolic product of T_3, also blocks the nongenomic actions of T_3 and T_4 that are initiated at the cell surface.

Endogenous or prescribed thyroid hormone appears to support tumor-relevant angiogenesis. It may also support pathologic skin neovascularization. In the absence of malignancy, however, the angiogenic activity of the hormone has exhibited experimental utility in wound-healing and tissue ischemia. Clinical applications of the hormone may be optimized by a nanoparticulate to which the hormone is covalently bound. Nanoparticulate thyroid hormone does not gain access to the intracellular space and acts exclusively at the hormone receptor on integrin $\alpha v \beta 3$. Restriction of the hormone to the extracellular space avoids unwanted hormonal effects on cellular respiration, on protein synthesis and on protein turnover and specific organ effects, such as cardiac tachyarrhythmias. Needed are more extensive preclinical data on the actions of thyroid hormone, T_4 and T_3 on, respectively, wound-healing and revascularization of limb or cardiac ischemic tissues.

References

1. Schmidt BM, Martin N, Georgens AC, Tillmann HC, Feuring M, Christ M, Wehling M (2002) Nongenomic cardiovascular effects of triiodothyronine in euthyroid male volunteers. J Clin Endocrinol Metab 87(4):1681–1686
2. Chilian WM, Wangler RD, Peters KG, Tomanek RJ, Marcus ML (1985) Thyroxine-induced left ventricular hypertrophy in the rat. Anatomical and physiological evidence for angiogenesis. Circ Res 57(4):591–598
3. Tomanek RJ, Busch TL (1998) Coordinated capillary and myocardial growth in response to thyroxine treatment. Anat Rec 251(1):44–49
4. Wang X, Zheng W, Christensen LP, Tomanek RJ (2003) DITPA stimulates bFGF, VEGF, angiopoietin, and Tie-2 and facilitates coronary arteriolar growth. Am J Physiol Circ Physiol 284(2):H613–H618
5. Davis FB, Mousa SA, O'Connor L, Mohamed S, Lin HY, Cao HJ, Davis PJ (2004) Proangiogenic action of thyroid hormone is fibroblast growth factor-dependent and is initiated at the cell surface. Circ Res 94(11):1500–1506
6. Bridoux A, Khan RA, Chen C, Cheve G, Cui H, Dyskin E, Yasri A, Mousa SA (2011) Design, synthesis and biological evaluation of bifunctional thyrointegrin inhibitors: new antiangiogenesis analogs. J Enzyme Inhib Med Chem 26(6):871–882
7. Mousa SA, Davis FB, Mohamed S, Davis PJ, Feng X (2006) Pro-angiogenesis action of thyroid hormone and analogs in a three-dimensional in vitro microvascular endothelial sprouting model. Int Angiol 25(4):407–413
8. Cheng SY, Leonard JL, Davis PJ (2010) Molecular aspects of thyroid hormone actions. Endocr Rev 31(2):139–170
9. Mylotte KM, Davis PJ, Davis FB, Blas SD, Schoenl M (1985) Milrinone and thyroid hormone stimulate myocardial membrane Ca^{2+}-ATPase activity and share structural homologies. Proc Natl Acad Sci USA 82(23):7974–7978
10. Davis FB, Davis PJ, Blas SD (1983) Role of calmodulin in thyroid hormone stimulation in vitro of human erythrocyte Ca^{2+}-ATPase activity. J Clin Invest 71(3):679–686
11. Rudinger A, Mylotte KM, Davis PJ, Davis FB, Blas SD (1984) Rabbit myocardial membrane Ca^{2+}-adenosine triphosphatase activity: stimulation in vitro by thyroid hormone. Arch Biochem Biophys 229(1):379–385
12. Incerpi S, Luly P, De Vito P, Farias RN (1999) Short-term effects of thyroid hormone on the Na/H antiport in L-6 myoblasts: high molecular specificity for 3, 3', 5-triiodo-L-thyronine. Endocrinology 140(2):683–689
13. Lei J, Mariash CN, Bhargava M, Wattenberg EV, Ingbar DH (2008) T3 increases Na-K-ATPase activity via a MAPK/ERK1/2-dependent pathway in rat adult alveolar epithelial cells. Am J Physiol Cell Molec Physiol 294(4):L749–L754
14. Segal J, Ingbar SH (1990) 3, 5, 3'-Triiodothyronine enhances sugar transport in rat thymocytes by increasing the intrinsic activity of the plasma membrane sugar transporter. J Endocrinol 124(1):133–140
15. Baumann CT, Maruvada P, Gl H, Yen PM (2001) Nuclear cytoplasmic shuttling by thyroid hormone receptors. Multiple protein interactions are required for nuclear retention. J Biol Chem 276(14):11237–11245
16. Sakaguchi Y, Cui G, Sen L (1996) Acute effects of thyroid hormone on inward rectifier potassium channel currents in guinea pig ventricular myocytes. Endocrinology 137(11):4744–4751

17. Davis PJ, Davis FB (2002) Nongenomic actions of thyroid hormone on the heart. Thyroid 12(6):459–466
18. Leonard JL, Farwell AP (1997) Thyroid hormone–regulated actin polymerization in brain. Thyroid 7(1):147–151
19. Siegrist-Kaiser CA, Juge-Aubry C, Tranter MP, Ekenbarger DM, Leonard JL (1990) Thyroxine-dependent modulation of actin polymerization in cultured astrocytes. A novel extranuclear action of thyroid hormone. J Biol Chem 265(9):5296–5302
20. Bergh JJ, Lin HY, Lansing L, Mohamed SN, Davis FB, Mousa S, Davis PJ (2005) Integrin alphav-beta3 contains a cell surface receptor for thyroid hormone that is linked to activation of mitogen-activated protein kinase and induction of angiogenesis. Endocrinology 146(7):2864–2871
21. Cody V, Davis PJ, Davis FB (2007) Molecular modeling of the thyroid hormone interaction with alpha v beta 3 integrin. Steroids 72(2):165–170
22. Lin HY, Cody V, Davis FB, Hercbergs A, Luidens MK, Mousa SA, Davis PJ (2011) Identification and functions of the plasma membrane receptor for thyroid hormone analogues. Discov Med 11(59):337–347
23. Plow EF, Haas TA, Zhang L, Loftus J, Smith JW (2000) Ligand binding to integrins. J Biol Chem 275(29):21785–21788
24. Davis PJ, Davis FB, Mousa SA, Luidens MK, Lin HY (2011) Membrane receptor for thyroid hormone: physiologic and pharmacologic implications. Annu Rev Pharmacol Toxicol 51:99–115
25. Lin HY, Sun M, Tang HY, Luidens MK, Mousa SA, Incerpi S, Drusano GL, Davis FB, Davis PJ (2009) L-Thyroxine vs. 3, 5, 3'-triiodo-L-thyronine and cell proliferation: activation of mitogen-activated protein kinase and phosphatidylinositol 3-kinase. Am J Physiol Cell Physiol 296(5):C980–C991
26. Glinskii AB, Glinsky GV, Lin HY, Tang HY, Sun M, Davis FB, Luidens MK, Mousa SA, Hercbergs AH, Davis PJ (2009) Modification of survival pathway gene expression in human breast cancer cells by tetraiodothyroacetic acid (tetrac). Cell Cycle 8(21):3554–3562
27. Lin HY, Lansing L, Merillon JM, Davis FB, Tang HY, Shih A, Vitrac X, Krisa S, Keating T, Cao HJ, Bergh J, Quackenbush S, Davis PJ (2006) Integrin alphavbeta3 contains a receptor site for resveratrol. FASEB J 20(10):1742–1744
28. Lin HY, Sun M, Lin C, Tang HY, London D, Ahih A, Davis FB, Davis PJ (2009) Androgen-induced human breast cancer cell proliferation is mediated by discrete mechanisms in estrogen receptor-alpha-positive and -negative breast cancer cells. J Steroid Biochem Mol Biol 113(305):182–188
29. Davis PJ, Davis FB, Mousa SA (2009) Thyroid hormone-induced angiogenesis. Curr Cardiol Rev 5(1):12–16
30. Luidens MK, Mousa SA, Davis FB, Lin HY, Davis PJ (2010) Thyroid hormone and angiogenesis. Vascul Pharmacol 52(3–4):142–145
31. Hsu AR, Veeravagu A, Hou LC, Tse V, Cher X (2007) Integrin alpha v beta 3 antagonists for anti-angiogenic cancer treatment. Recent Pat Anticancer Drug Discov 2(2):143–158
32. Gaertner FC, Schwaiger M, Beer AJ (2010) Molecular imaging of $\alpha v\beta 3$ expression in cancer patients. Q J Nucl Med Mol Imaging 54(3):309–326
33. Lu X, Wang RF (2012) A concise review of current radiopharmaceuticals in tumor angiogenesis imaging. Curr Pharm Des 18(8):1032–1040
34. Mousa SS, Davis FB, Davis PJ, Mousa SA (2010) Human platelet aggregations and degranulation is induced in vitro by L-thyroxine, but not by 3, 5, 3'-triiodo-L-thyronine or diiodothyropropionic acid (DITPA). Clin Appl Thromb Hemost 16(3):288–293
35. Hoffman SJ, Vasko-Moser J, Miller WH, Lark MW, Gowen M, Stroup G (2002) Rapid inhibition of thyroxine-induced bone resorption in the rat by an orally active vitronectin receptor antagonist. J Pharmacol Exp Ther 302(1):205–211
36. El-Eter E, Rebbaa H, Alkayali A, Mousa SA (2007) Role of thyroid hormone analogues in angiogenesis and the development of collaterals in the rabbit hind limb ischemia model. J Thrombo Thrombolysis 5(suppl 1):375
37. Mousa SA, Bergh JJ, Dier E, Rebbaa A, O'Connor LJ, Yalcin M, Aljada A, Dyskin E, Davis FB, Lin HY, Davis PJ (2008) Tetraiodothyroacetic acid, a small molecule integrin ligand,

blocks angiogenesis induced by vascular endothelial growth factor and basic fibroblast growth factor. Angiogenesis 11(2):183–190

38. Borges E, Jan Y, Ruoslahti E (2000) Platelet-derived growth factor receptor beta and vascular endothelial growth factor receptor 2 bind to the beta3 integrin through its extracellular domain. J Biol Chem 275(51):39867–39873

39. Somananth PR, Malinin NL, Byzova TV (2009) Cooperation between integrin alphavbeta3 and VEGFR2 in angiogenesis. Angiogenesis 12(2):177–185

40. Mousa SA, O'Connor L, Davis FB, Davis PJ (2006) Proangiogenic action of the thyroid hormone analog 3, 5-diiodothyropionic acid (DITPA) is initiated at the cell surface and is integrin-mediated. Endocrinology 147(4):1602–1607

41. Mousa SA, O'Connor L, Bergh JJ, Davis FB, Scanlan TS, Davis PJ (2005) The proangiogenic action of thyroid hormone analogue GC-1 is initiated at an integrin. J Cardiovasc Pharmacol 46(3):356–360

42. Bharali DJ, Yalcin M, Davis PJ, Mousa SA (2013) Tetraiodothyroacetic acid-conjugated poly(lactic-co-glycolic acid) nanoparticles: a nanomedicine approach to treat drug-resistant breast cancer. Nanomedicine (London). doi:10.2217/NNM.12.200

43. Bridoux A, Cui H, Dyskin E, Yalcin M, Mousa SA (2009) Semisynthesis and pharmacological activities of tetrac analogs: angiogenesis modulators. Bioorg Med Chem Lett 19(12):3259–3263

44. Yalcin M, Dyskin E, Lansing L, Bharali DJ, Mousa SS, Bridoux A, Hercbergs AH, Lin HY, Davis FB, Glinsky GV, Glinskii A, Ma J, Davis PJ, Mousa SA (2010) Tetraiodothyroacetic acid (tetrac) and nanoparticulate tetrac arrest growth of medullary carcinoma of the thyroid. J Clin Endocrinol Metab 95(4):1972–1980

45. Yalcin M, Bharali DJ, Lansing L, Dyskin E, Mousa SS, Hercbergs A, Davis FB, Davis PJ, Mousa SA (2009) Tetraiodothyroacetic acid (tetrac) and tetrac nanoparticles inhibit growth of human renal cell carcinoma xenografts. Anticancer Res 29(10):3825–3831

46. Yalcin M, Bharali DJ, Dyskin E, Dier E, Lansing L, Mousa SS, Davis FB, Davis PJ, Mousa SA (2010) Tetraiodothyroacetic acid and tetraiodothyroacetic acid nanoparticle effectively inhibit the growth of human follicular thyroid cell carcinoma. Thyroid 20(3):281–286

47. Mousa SA, Yalcin M, Bharali DJ, Meng R, Tang HY, Lin HY, Davis FB, Davis PJ (2012) Tetraiodothyroacetic acid and its nanoformulation inhibit thyroid hormone stimulation of non-small cell lung cancer cells in vitro and its growth in xenografts. Lung Cancer 76(1):39–45

48. Strickler JH, Hurwitz HI (2012) Bevacizumab-based therapies in the first-line treatment of metastatic colorectal cancer. Oncologist 17(4):513–524

49. Shojaei F (2012) Anti-angiogenesis therapy in cancer: current challenges and future perspectives. Cancer Lett 320(2):130–137

50. Aggarwal C, Somaiah N, Simon G (2012) Antiangiogenic agents in the management of non-small cell lung cancer: where do we stand now and where are we headed? Cancer Biol Ther 13(5):247–263

51. Meng R, Tang HY, Westfall J, London D, Cao JH, Mousa SA, Luidens M, Hercbergs A, Davis FB, Davis PJ, Lin HY (2011) Crosstalk between integrin $\alpha v \beta 3$ and estrogen receptor-α is involved in thyroid hormone-induced proliferation in human lung carcinoma cells. PLoS One 6(11):e27547

52. Lin HY, Tang HY, Shih A, Keating T, Cao G, Davis PJ, Davis FB (2007) Thyroid hormone is a MAPK-dependent growth factor for thyroid cancer cells and is anti-apoptotic. Steroids 72(2):180–187

53. Davis FB, Tang HY, Shih A, Keating T, Lansing L, Hercbergs A, Fenstermaker RA, Mousa A, Mousa SA, Davis PJ, Lin HY (2006) Acting via a cell surface receptor, thyroid hormone is a growth factor for glioma cells. Cancer Res 66(14):7270–7275

54. Hercbergs AA, Goyal LK, Suh JH, Lee S, Reddy CA, Cohen BH, Stevens GH, Reddy SK, Peereboom DM, Elson PJ, Gupta MK, Barnett GH (2003) Propylthiouracil-induced chemical hypothyroidism with high-dose tamoxifen prolongs survival in recurrent high grade glioma: a phase I/II study. Anticancer Res 23(1B):617–626

55. Cristofanilli M, Yamamura Y, Kau SW, Bevers T, Strom S, Patangan M, Hsu L, Krishnamurthy S, Theriault RL, Hortobagyi GN (2005) Thyroid hormone and breast carcinoma. Primary hypothyroidism is associated with a reduced incidence of primary breast carcinoma. Cancer 103(6):1122–1128

56. Riesenbeck LM, Bierer S, Hoffmeister J, Kopke T, Papavasilis P, Hertle L, Thielen B, Herrmann E (2011) Hypothyroidism correlates with a better prognosis in metastatic renal cancer patients treated with sorafenib or sunitinib. World J Urol 29(6):807–813

57. Schmidinger M, Vogl UM, Bojic M, Lamm W, Heinzl H, Haitel A, Clodi M, Kramer G, Zielinski CC (2011) Hypothyroidism in patients with renal cell carcinoma: blessing or curse? Cancer 117(3):534–544

58. Hercbergs AH, Ashur-Fabian O, Garfield D (2011) Thyroid hormones and cancer: clinical studies of hypothyroidism in oncology. Curr Opinion Endocrinol Diabetes Obes 17(5):432–436

59. Turkyilmaz A, Eroglu A, Avdin Y, Yilmaz O, Karaoglanoglu N (2010) A new risk factor in oesophageal cancer aetiology: hyperthyroidism. Acta Chir Belg 110(5):533–536

60. Tosovic A, Becker C, Bondeson AG, Bondeson L, Ericsson UB, Malm J, Manjer J (2012) Prospectively measured thyroid hormones and thyroid peroxidase antibodies in relation to breast cancer risk. Int J Cancer 131(9):2126–2133

61. Tomanek RJ, Zimmerman MB, Suvama PR, Morkin E, Pennock GD, Goldman S (1998) A thyroid hormone analog stimulates angiogenesis in the post-infarcted rat heart. J Mol Cell Cardiol 30(5):923–932

62. Zheng W, Weiss RM, Wang X, Zhou R, Arlen AM, Lei L, Lazartiques E, Tomanek RJ (2004) DITPA stimulates arteriolar growth and modifies myocardial postinfarction remodeling. Am J Physiol Circ Physiol 286(5):H1994–H2000

63. Savinova OV, Liu Y, Aasen GA, Mao K, Weltman NY, Nedich BL, Liang Q, Gerdes AM (2011) Thyroid hormone promotes remodeling of coronary resistance vessels. PLoS One 6(9):e25054

64. Natori J, Shimizu K, Nagahama M, Tanaka S (1999) The influence of hypothyroidism on wound healing. An experimental study. Nihon Ika Daigaku Zasshi 66(3):176–180

65. Safer JD, Crawford TM, Holick MF (2005) Topical thyroid hormone accelerates wound healing in mice. Endocrinology 146(10):4425–4430

66. Safer JD, Crawford TM, Holick MF (2004) A role for thyroid hormone in wound healing through keratin gene expression. Endocrinology 145(5):2357–2361

67. Tha Nassif AC, Hintz Greca F, Graf H, Domingues Repka JC, Nassif LS (2009) Wound healing in colonic anastomosis in hypothyroidism. Eur Surg Res 42(4):209–215

68. Zimmermann E, Ribas-Filho JM, Malafaia O, Ribas CA, Nassif PA, Stieven Filho E, Przysiezny PE (2009) Tracheal suture in rats with hypothyroidism: wound healing study. Acta Cir Bras 24(4):282–289

69. Ozawa J, Kawamata S, Kurosaki T, Iwamizu Y, Matsuura N, Abiko S, Kai S (2003) Ischemic injury and repair process after transection in hypothyroid rat muscles. Muscle Nerve 27(5):595–603

70. Zhang L, Cooper-Kuhn CM, Nannmark U, Blomgren K, Kuhn HG (2010) Stimulatory effects of thyroid on brain angiogenesis in vivo and in vitro. J Cereb Blood Flow Metab 30(2):323–335

71. Schlenker EH, Hora M, Liu Y, Redetzke RA, Morkin E, Gerdes AM (2008) Effects of thyroidectomy, T4, and DITPA replacement on brain blood vessel density in adult rats. Am J Physiol Regul Integr Comp Physiol 294(5):R1504–R1509

72. Jiang JY, Miyabayashi K, Nottola SA, Umezu M, Cecconi S, Sato E, Macchiarelli G (2008) Thyroxine treatment stimulated ovarian follicular angiogenesis in immature hypothyroid rats. Histol Histopathol 23(11):1387–1398

73. Zhao W, Gao BL, Liu ZY, Yi GF, Shen LJ, Yang HY, Li H, Song DP, Jiang YN, Hu JH, Luo G (2009) Angiogenic study in Graves' disease treated with thyroid arterial embolization. Clin Invest Med 32(5):E335–E344

74. Tseleni-Balafouta S, Kavantzas N, Balafoutas D, Patsouris E (2006) Comparative study of angiogenesis in thyroid glands with Graves disease and Hashimoto's thyroiditis. Appl Immunohistochem Mol Morphol 14(2):203–207
75. Kim BY, Kim CH, Jung CH, Mok JO, Suh KI, Kang SK (2011) Association between subclinical hypothyroidism and severe diabetic retinopathy in Korean patients with type 2 diabetes. Endocr J 58(12):1065–1070
76. Yang GR, Yang JK, Zhang L, An YH, Lu JK (2010) Association between subclinical hypothyroidism and proliferative diabetic retinopathy: a case–control study. Tohoku J Exp Med 222(4):303–310
77. Yang JK, Liu W, Shi J, Li YB (2010) An association between subclinical hypothyroidism and sight-threatening diabetic retinopathy in type 2 diabetic patients. Diabetes Care 33(5):1018–1020
78. Soh EY, Sobhi SA, Wong MG, Meng YG, Siperstain AE, Clark OH, Duh QY (1996) Thyroid stimulating hormone promotes the secretion of VEGF in thyroid cancer cell lines. Surgery 120(6):944–947
79. Klein M, Brunaud L, Muresan M, Barbe F, Marie B, Spain R, Vignaud JM, Chatelin J, Angiol-Duprez K, Zamegar R, Weryha G, Duprez A (2006) Recombinant human thyrotropin stimulates thyroid angiogenesis in vivo. Thyroid 16(6):531–536
80. Yoshida T, Gong J, Xu Z, Wei Y, Duh EJ (2012) Inhibition of pathological angiogenesis by the integrin $\alpha v\beta 3$ antagonist tetraiodothyroacetic acid (tetrac). Exp Eye Res 94(1):41–48
81. Smith JR, Lanier VB, Braziel RM, Falkenhagen KM, White C, Rosenbaum JT (2007) Expression of vascular endothelial growth factor and its receptors in rosacea. Br J Ophthalmol 91(2):226–229
82. Kusumanto YH, Dam WA, Hospers GA, Meijer C, Mulder NH (2003) Platelets and granulocytes, in particular the neutrophils, form important compartments for circulating vascular endothelial growth factor. Angiogenesis 6(4):283–287
83. Yamasaki E, Soma Y, Nakamura M, Kawa Y, Mizoguchi M (2004) Differential regulation of the secretion of vascular endothelial growth factor from human skin fibroblasts by growth factors and cytokines. J Dermatol Sci 36(2):118–121
84. Yamagata M, Mikami T, Tsurula T, Yokoyama K, Sada M, Kobayaashi K, Katsumata T, Koizumi W, Saigenji K, Okayasu I (2011) Submucosal fibrosis and basic-fibroblast growth factor-positive neutrophils correlate with colonic stenosis in cases of ulcerative colitis. Digestion 84(1):12–21
85. Yaguchi H, Tsuboi R, Ueki R, Ogawa H (1993) Immunohistochemical localization of basic fibroblast growth factor in skin diseases. Acta Derm Venereol 73(2):81–83
86. Shih A, Zhang S, Cao HJ, Tang HY, Davis FB, Davis PJ, Lin HY (2004) Disparate effects of thyroid hormone on actions of epidermal growth factor and transforming growth factor-alpha are mediated by 3', 5'-cyclic adenosine 5'-monophosphate-dependent protein kinase II. Endocrinology 145(4):1708–1717
87. Iivanainen E, Elenius K (2010) ErbB targeted drugs and angiogenesis. Curr Vasc Pharmacol 8(3):421–431
88. Pavolov KA, Gershtein ES, Dubova EA, Shchegolev AI (2011) Vascular endothelial growth factor and type 2 receptor for this factor in vascular malformations. Bull Exp Biol Med 150(4):481–484
89. Weinsheimer SM, Xu H, Achrol AS, Stamova B, McCulloch C, Pawlikowska L, Tian Y, Ko NU, Lawton MT, Steinberg GK, Chang SD, Jickling G, Ander BP, Kim H, Sharp FR, Young WL (2011) Gene expression profiling of blood in brain arteriovenous malformation patients. Transl Stroke Res 2(4):575–587
90. Koenig RJ (2008) Modeling the nonthyroidal illness syndrome. Curr Opin Endocrinol Diabetes Obes 15(5):466–469
91. Chen YF, Redetzke RA, Said S, Beyer AJ, Gerdes AM (2010) Changes in left ventricular function and remodeling after myocardial infarction in hypothyroid rats. Am J Physiol Circ Physiol 298(1):H259–H262

Chapter 5
Actions of Steroids and Peptide Hormones on Angiogenesis

Paul J. Davis, Shaker A. Mousa, Faith B. Davis, and Hung-Yun Lin

Abstract Levels of normal or pathologic angiogenic activity are usually considered to be products of contributions from principal vascular growth factors and their specific receptors, of anatomic locations in specific organs or, particularly in cancer, of tissue oxygen tension and of pharmaceutical modulation. In this chapter, we examine the contributions of endocrine hormones—small molecules (steroids) and peptides—to regulation of angiogenesis. Most of the hormones considered appear to have actions on blood vessels only in specific tissues, but the angiogenic properties of these substances have not been systematically studied in tissues other than those that are classical targets. Estrogens and androgens have angiogenic actions, some of which are obtained exclusively in cells obtained, respectively, from female and male sources. Estrogen and progesterone contribute to uterine angiogenesis in the menstrual cycle. Glucocorticoids that lack classical glucocorticoid activities have been designed pharmaceutically to be angiostatic drugs. Among the peptide hormones, prolactin is angiogenic in certain tissues, but the hormone is proteolytically cleaved to yield angiostatic peptides. ACTH of course affects angiogenesis in the adrenal cortex, but the ACTH receptor may be expressed in other tissues, e.g., the placenta, which then may be the focus of ACTH-dependent angiogenesis. TSH is pro-angiogenic in the thyroid gland and supports tumor-related carcinoma of the thyroid, but

P.J. Davis (✉)
Department of Medicine, Albany Medical Center, Albany, NY, USA

The Pharmaceutical Research Institute at Albany College of Pharmacy
and Health Sciences, Rensselaer, NY, USA
e-mail: pdavis.ordwayst@gmail.com

S.A. Mousa • F.B. Davis
The Pharmaceutical Research Institute, Albany College of Pharmacy
and Health Sciences, Rensselaer, NY, USA

H.-Y. Lin
Institute of Cancer Biology and Drug Discovery, Taipei Medical University, Taipei, Taiwan

S.A. Mousa and P.J. Davis (eds.), *Angiogenesis Modulations in Health and Disease:*
Practical Applications of Pro- and Anti-angiogenesis Targets,
DOI 10.1007/978-94-007-6467-5_5, © Springer Science+Business Media Dordrecht 2013

the TSH receptor may be expressed in other tissues and, in such settings, circulating levels of TSH may be supporting new blood vessel formation. In a separate chapter we have described the angiogenic spectrum of thyroid hormone and its analogues.

The effects of thyroid hormone and thyroid hormone analogues on angiogenesis are reviewed in this textbook in Chap. 4. These actions are initiated at a cell surface receptor on integrin αvβ3, and much of the downstream mechanism underlying the effects on blood vessels has been elucidated [1]. In this chapter, we examine other small molecule hormones—the steroids—for actions on neovascularization and also look at effects of peptide hormones on angiogenesis.

Steroid Hormones

Estrogen

Via a nongenomic mechanism, estrogen has a variety of effects on endothelial cells and angiogenesis [2]. X-ray crystallographic modeling of integrin αvβ3 suggests the presence of an estrogen-binding site [3] that would be a candidate initiation site for pro-angiogenic action of estrogen.

Estradiol has activity as an inducer in women of proliferation of human endothelial progenitor cells (EPCs) harvested from peripheral blood [4]. This action may be genomic, i.e., dependent upon a nuclear estrogen receptor protein, specifically, ERα. Interestingly, the effect depends upon the menstrual cycle phase of the donor, being absent in the luteal phase.

Progesterone

Progesterone, like estrogen, is angiogenic and is capable of stimulating human peripheral blood EPC proliferation [4]. Its effects on neovascularizatrion in the uterus during the implantation phase of pregnancy have been shown to involve expression of a specific angiopoietin-like gene (angiopoietin-like 4, Angptl4) [5].

Androgen

The contribution of testosterone to post-myocardial infarction revascularization in the heart has been described [6], as have other pro-angiogenic actions of this steroid. The molecular mechanism may be assumed to involve the nuclear receptor for testosterone (androgen receptor, AR).

Testosterone and dihydrotestosterone also promote vascular tube formation in the Matrigel® assay [7]. The pro-angiogenic activity of androgens is demonstrable in male-source endothelial cells, but not in female-source cells [8]. Others have also shown the sexual dimorphism of androgen effects on endothelial cells [9]. Overexpression of nuclear androgen receptor (AR) in female endothelial cells confers testosterone sensitivity in angiogenic assays [8]. Thus, these endothelial cell actions of testosterone are genomic in mechanism.

Nongenomic mechanisms may also be involved in androgen-directed angiogenesis. A plasma membrane receptor site for testosterone has been reported that is involved in tumor cell proliferation and that may be a candidate initiation site for other nongenomic blood vessel actions of the hormone [10]. Further, the increase in expression of vasostimulatory *hypoxia-inducible factor 1α (HIF-1α)* in response to testosterone [6] we know can be induced nongenomically—that is, independently of nuclear hormone receptors—by hormones other than androgens [11].

Adrenal Corticoids

High-dose glucocorticoids are angiostatic and may potentiate anti-angiogenic properties of other agents. Dexamethasone, for example, enhanced the angiostatic activity of cisplatin in Ehrlich ascites carcinoma in mice [12]. Corticosterone inhibits angiogenesis in muscle [13] and in specific regions of rat brain [14]. More than a decade ago, McNatt et al. [15] compared the anti-angiogenic effects of 19 corticosteroids in the CAM assay, and the structures of some of the proprietary agents studied have led to interests in design of angiostatic glucocorticoids that have no traditional glucocorticoid effects and are designated 'cortisenes.' Most of these drugs—such as anecortave acetate [16] and derivatives [17]—are intended for use against intraocular vascular disease, such as age-related macular degeneration. It is not yet clear whether such agents may be practical in a systemic context, such as tumor-related angiogenesis.

Peptide Hormones

Prolactin (PRL)

PRL induces tube formation by endothelial cells in the Matrigel® assay [18] and is pro-angiogenic in the CAM assay. There is little or no action on endothelial cell proliferation, but migration of these cells is stimulated. The PRL receptor is identifiable immunohistochemically in the microvasculature of human breast cancer [18].

The proteolysis of prolactin yields a set of peptides, the vasoinhibins, that are anti-angiogenic [19] and arrest tumor growth. Thus, local regulation of proteases involved in the cleavage of PRL is a potential target for angiogenesis regulation in, for example, breast cancer. In addition to its angiostatic activity, the 16 kDa N-terminal fragment of human PRL (16K hPRL) is a lymphangiostatic agent that reduces abundance of lymphatics in primary tumor and in sentinel lymph nodes [20]

Adrenocorticotrophic Hormone (ACTH)

The action of ACTH on adrenal cortical cells and on steroidogenesis is complimented by actions of the hormone on angiogenesis in the adrenal cortex [21]. The mechanism of this action involves increased mRNA abundance of multiple VEGF isoforms [21] and, at least in the fetal adrenal cortex, of angiopoietin 2 (Ang-2) [22], a factor that destabilizes blood vessel structure, rendering it more responsive to angiogenic stimuli such as VEGF. The possibility that ACTH may support angiogenesis in organs other than the adrenal cortex has not been examined, although the receptor has been described in other tissues, such as the placenta [23].

Luteinizing Hormone (LH)

LH is pro-angiogenic in the uterus during early pregnancy. The mechanism in stromal cells exposed to LH is VEGF-dependent [24].

Elevated circulating gonadotropin levels are a risk factor for ovarian cancer development. The action of LH on angiogenesis that supports ovarian cancer growth has been found to depend upon up-regulation of VEGF production and of a receptor for VEGF [25]. The biochemical mechanism of this adjustment has been found to involve the phosphatidylinositol 3-kinase (PI3K) signal transduction pathway and to be subject to inhibition by the biguanide, metformin [25].

Follicle-stimulating Hormone (FSH)

During ovarian follicle development, FSH is pro-angiogenic in granulosa cells. The mechanism of this effect has been partially elucidated and involves increased local expression of VEGF and platelet-derived growth factor subunit B (PDGF-B) [26]. Transforming growth factor β1 (TGF-β1) facilitates differentiation of granulosa cells induced by FSH, and TGF-β1, itself, up-regulates local secretion of VEGF and PDGF-B [26].

Thyrotropin (TSH)

TSH stimulates normal thyroid gland and thyroid tumor growth. The growth requires angiogenesis, and TSH has been shown to increase VEGF production at the level of transcription in thyroid cancer cells [27]. The mechanism is protein kinase C (PKC)-dependent. At the same time and paradoxically, TSH may increase endostatin production by certain thyroid tumor cell lines, but not in others [28], regardless of the expression of the TSH receptor by these cells.

The TSH receptor (TSHR) may be ectopically expressed in nonthyroidal tissues, e.g., adrenal cells [29] and in osteosarcoma cells [30], raising the possibility that circulating TSH may be an angiogenic factor in certain nonthyroidal tissues. Thus, in the setting of primary hypothyroidism when endogenous, pro-angiogenic L-thyroxine (T_4) and 3, 5, 3'-triiodo-L-thyronine (T_3) [1, 31] are decreased, the elevated serum TSH that is present might support blood vessel formation in the rare circumstance of ectopic TSHR production.

Insulin

Insulin promotes angiogenesis, at least in part via local VEGF production [32] and stimulation of endothelial cell proliferation. Disordering of the PI3K and mitogen-activated protein kinase (MAPK) signal transduction pathways occurs in insulin resistance [33]. This may explain the impaired angiogenesis that is a part of this clinical state since these pathways contribute to endothelial cell proliferation induced by insulin.

Melatonin

Melatonin can interfere with estrogen action on breast cancer cells and is also anti-angiogenic in tumors. Melatonin inhibits VEGF production in human breast cancer MCF-7 cells and blocks proliferation of human umbilical vein endothelial cells co-cultured with MCF-7 cells [34]. It is not yet clear in what other tumoral or nonmalignant cells that melatonin may affect transcription of the *VEGF* gene.

Conclusions

A spectrum of endocrine hormones—both small molecules and peptide hormones—is described here to have distinctive experimental effects on new blood vessel formation. How important such effects may be clinically has yet in most cases to be clearly characterized. Where a clinical impact may be defined, it will depend upon the secretion pattern of the hormone, pathologic elevations or losses of the hormones, the normal or tumoral tissue targets of the hormones and the presence of other factors in blood or tissue that are relevant to angiogenesis. Another perspective from which to look at the substantial number of endocrine hormones with pro- or anti-angiogenic properties is that of an algebraic sum of the activities of these factors affecting basal levels of blood vessel formation. Some of the hormones reviewed are or may be models for pharmaceutical development.

References

1. Davis PJ, Davis FB, Mousa SA, Luidens MK, Lin HY (2011) Membrane receptor for thyroid hormone: physiologic and pharmacologic implications. Annu Rev Pharmacol Toxicol 51:99–115
2. Kim KH, Bender JR (2009) Membrane-initiated actions of estrogen on the endometrium. Mol Cell Endocrinol 308(1–2):3–8
3. Lin HY, Cody V, Davis FB, Hercbergs AA, Luidens MK, Mousa SA, Davis PJ (2011) Identification and functions of the plasma membrane receptor for thyroid hormone analogues. Discov Med 11(59):337–347
4. Matsubara Y, Matsubara K (2012) Estrogen and progesterone play pivotal roles in endothelial progenitor cell proliferation. Reprod Biol Endocrinol 10:2
5. Scott CA, van Huyen D, Bany BM (2012) Angiopoietin-like gene expression in the mouse uterus during implantation and in response to steroids. Cell Tissue Res 348(1):199–211
6. Chen Y, Fu L, Han Y, Teng Y, Sun J, Xie R, Cao J (2012) Testosterone replacement therapy promotes angiogenesis after acute myocardial infarction by enhancing expression of cytokines HIF-1α, SDF-1α and VEGF. Eur J Pharmacol 684(1–3):116–124
7. Liao H, Zhou Q, Gu Y, Duan T, Feng Y (2012) Luteinizing hormone facilitates angiogenesis in ovarian epithelial tumor cells and metformin inhibits the effect through the mTOR signaling pathway. Oncol Rep 27(6):1873–1878
8. Sieveking DP, Lim P, Chow RWY, Dunn LL, Bao S, McGrath KCY, Heather AK, Handelsman DJ, Celemajer DS, Ng MKC (2010) A sex-specific role for androgens in angiogenesis. J Exp Med 207(2):345–352
9. Rubinow KB, Amory JK, Page ST (2011) Androgens exert sexually dimorphic effects on angiogenesis: novel insight into the relationship between androgens and cardiovascular disease. Asian J Androl 13(4):626–627
10. Lin HY, Sun M, Lin C, Tang HY, London D, Shih A, Davis FB, Davis PJ (2009) Androgen-induced human breast cancer cell proliferation is mediated by discrete mechanisms in estrogen receptor-α-positive and -negative breast cancer cells. J Steroid Biochem Mol Biol 113(3–5):182–188
11. Lin HY, Sun M, Tang HY, Lin C, Luidens MK, Mousa SA, Inserpi S, Drusano GL, Davis FB, Davis PJ (2009) L-Thyroxine vs. 3, 5, 3'-triiodo-L-thyronine and cell proliferation: activation of mitogen-activated protein kinase and phosphatidylinositol 3-kinase. Am J Physiol Cell Physiol 296(5):C980–C991
12. Arafa HM, Abdel-Hamid MA, El-Khouly AA, Elmazar MM, Osman AM (2006) Enhancement by dexamethasone of the therapeutic benefits of cisplatin via regulation of tumor angiogenesis and cell cycle kinetics in a murine tumor paradigm. Toxicology 222(1–2):103–113
13. Shikatani EA, Trifonova A, Mandel ER, Liu ST, Roudier E, Krylova A, Szigiato A, Beaudry J, Riddell MC, Haas TL (2012) Inhibition of proliferation, migration and proteolysis contribute to corticosterone-mediated inhibition of angiogenesis. PLoS One 7(10):e46625
14. Ekstrand J, Hellsten J, Tingstrom A (2008) Environmental enrichment, exercise and corticosterone affect endothelial cell proliferation in adult rat hippocampus and prefrontal cortex. Neurosci Lett 442(3):203–207
15. McNatt LG, Weimer L, Yanni J, Clark AF (1999) Angiostatic activity of steroids in the chick embryo CAM and rabbit cornea models of neovascularization. J Ocul Pharmacol Ther 15:413–423
16. Schmidt-Erfurth U, Michels S, Michels R, Aue A (2005) Anecortave acetate for the treatment of subfoveal choroidal neovascularization secondary to age-related macular degeneration. Eur J Ophthalmol 15(4):482–485
17. Missel P, Chastain J, Mitra A, Kompella U, Kansara V, Duvvuri S, Amrite A, Cheruvu N (2010) In vitro transport and partitioning of AL-4940, active metabolite of angiostatic agent anecortave acetate, in ocular tissues of the posterior segment. J Ocul Pharmacol Ther 26(2):137–146

18. Reuwer AQ, Nowak-Sliwinska P, Mans LA, van der Loos CM, von der ThusenJH TMT, Spek CA, Goffin V, Griffioen AW, Borensztajn KS (2012) Functional consequences of prolactin signaling on endothelial cells: a potential link with angiogenesis in pathophysiology? J Cell Mol Med 16(9):2035–2048
19. Clapp C, Martinez de la Escalera L, Martinez de la Escalera G (2012) Prolactin and blood vessels: a comparative endocrinology perspective. Gen Comp Endocrinol 176(3):336–340
20. Kinet V, Castermans K, Herkenne S, Maillard C, Blacher S, Lion M, Noel A, Martial JA, Struman I (2011) The angiostatic protein 16K human prolaction significantly prevents tumor-induced lymphangiogenesis by affecting lymphatic endothelial cells. Endocrinology 152(11):4062–4071
21. Mallet C, Feraud O, Ouengue-Mbele G, Gaillard I, Sappay N, Vittet D, Vilgrain I (2003) Differential expression of VEGF receptors in adrenal atrophy induced by dexamethasone: a protective role of ACTH. Am J Physiol Endocrinol Metab 284(1):E156–E167
22. Ishimoto H, Ginzinger DG, Jaffe RB (2006) Adrenocorticotropin preferentially up-regulates angiopoietin 2 in the human fetal adrenal gland: implications for coordinated adrenal organ growth and angiogenesis. J Clin Endocrinol Metab 91(5):1909–1915
23. Izumi S, Abe K, Hayashi T, Nakane PK, Koji T (2004) Immunohistochemical localization of the ACTH (MC-2) receptor in the rat placenta and adrenal gland. Arch Histol Cytol 67(5):443–453
24. Kaczmarek MM, Blitek A, Schams D, Ziecik AJ (2010) Effect of luteinizing hormone and tumour necrosis factor-α on VEGF secretion by cultured porcine endometrial stromal cells. Reprod Domest Anim 45(3):481–486
25. Liao CH, Lin FY, Wu YN, Chiang HS (2012) Androgens inhibit tumor necrosis factor-α-induced cell adhesion and promote tube formation of human coronary artery endothelial cells. Steroids 77(7):756–764
26. Kuo SW, Ke FC, Chang GD, Lee MT, Hwang JJ (2011) Potential role of follicle-stimulating hormone (FSH) and transforming growth factor (TGFβ1) in the regulation of ovarian angiogenesis. J Cell Physiol 226(6):1608–1619
27. Hoffmann S, Hofbauer LC, Scharrenbach V, Wunderlich A, Hassan I, Lingelbach S, Zielke A (2004) Thyrotropin (TSH)-induced production of vascular endothelial growth factor in thyroid cancer cells in vitro: evaluation of TSH signal transduction and of angiogenesis-stimulating growth factors. J Clin Endocrinol Metab 89(12):6139–6145
28. Hoffman S, Wunderlich A, Lingelbach S, Musholt PB, Musholt TJ, von Wasielewski R, Zielke A (2008) Expression and secretion of endostatin in thyroid cancer. Ann Surg Oncol 15(12):3601–3608
29. Zwermann O, Suttmann Y, Bidlingmaier M, Beuschlein F, Reincke M (2009) Screening for membrane hormone receptor expression in primary aldosteronism. Eur J Endocrinol 160(3):443–451
30. Inoue M, Tawata M, Yokomori N, Endo T, Onaya T (1998) Expression of thyrotropin receptor on clonal osteoblast-like rat osteosarcoma cells. Thyroid 8(11):1059–1064
31. Cheng SY, Leonard JL, Davis PJ (2010) Molecular aspects of thyroid hormone actions. Endocr Rev 31(2):139–170
32. Iliadis F, Kadoglou N, Didangelos T (2011) Insulin and the heart. Diabetes Res Clin Pract 93(Suppl 1):S86–S91
33. Cubbon RM, Ali N, Sengupta A, Kearney MT (2012) Insulin- and growth factor-resistance impairs vascular regeneration in diabetes mellitus. Curr Vasc Pharmacol 10(3):271–284
34. Alvarez-Garcia V, Gonzalez A, Alonso-Gonzalez C, Martinez-Campa C, Cos S (2012) Regulation of vascular endothelial growth factor by melatonin in human breast cancer cells. J Pineal Res. 54(4):373–380

Chapter 6
Role of Non-neuronal Nicotinic Acetylcholine Receptors in Angiogenesis Modulation

Shaker A. Mousa, Hugo R. Arias, and Paul J. Davis

Abstract Angiogenesis is a critical physiological process for cell survival and development. Endothelial cells, necessary for the course of angiogenesis, express several non-neuronal nicotinic acetylcholine receptors (AChRs). The most important functional non-neuronal AChRs are homomeric α7 AChRs and several heteromeric AChRs formed by a combination of α3, α5, β2, and β4 subunits, including α3β4-containing AChRs. In endothelial cells, α7 AChR stimulation indirectly triggers the activation of the integrin αvβ3 receptor and an intracellular MAP kinase (ERK) pathway that mediates angiogenesis. Non-selective cholinergic agonists such as nicotine have been shown to induce angiogenesis, enhancing tumor progression. Moreover, α7 AChR selective antagonists such as α-bungarotoxin and methyllycaconitine as well as the nonspecific antagonist mecamylamine have been shown to inhibit endothelial cell proliferation and ultimately blood vessel formation. Exploitation of such pharmacologic properties can lead to the discovery of new specific cholinergic antagonists as anti-cancer therapies. Conversely, the proangiogenic effect elicited by specific agonists can be used to treat diseases that respond to revascularization such as diabetic ischemia and foot ulcer, as well as to

S.A. Mousa (✉)
The Pharmaceutical Research Institute at Albany College of Pharmacy
and Health Sciences, Rensselaer, NY, USA
e-mail: shaker.mousa@acphs.edu

H.R. Arias
Department of Medical Education, College of Medicine, California North State University,
Elk Grove, CA, USA

P.J. Davis
The Pharmaceutical Research Institute at Albany College of Pharmacy
and Health Sciences Albany, Rensselaer, NY, USA

Department of Medicine, Albany Medical Center, Albany, NY, USA
e-mail: pdavis.ordwayst@gmail.com

S.A. Mousa and P.J. Davis (eds.), *Angiogenesis Modulation in Health and Disease:*   55
Practical Applications of Pro- and Anti-angiogenesis Targets,
DOI 10.1007/978-94-007-6467-5_6, © Springer Science+Business Media Dordrecht 2013

accelerate wound-healing. In this chapter we discuss the pharmacological evidence supporting the importance of non-neuronal AChRs in angiogenesis. We also explore potential intracellular mechanisms by which α7 AChR activation mediates this vital cellular process.

Abbreviations

α-BTx	α-bungarotoxin
ACh	acetylcholine
AChR	nicotinic acetylcholine receptor
Akt	protein kinase B
BDNF	brain-derived neurotrophic factor
bFGF (FGF2)	basic fibroblast growth factor
CAM	chorioallantoic membrane
COX-2	cyclooxygenase-2
DHβE	dihydro-β-erythroidine
DMPP	dimethylphenylpiperazinium
EC	endothelial cell
EGFR	epidermal growth factor receptor
ERK	extracellular signal-regulated kinase
FGF	fibroblast growth factor
FLT1	FMS-related tyrosine kinase-1
HIF-1	hypoxia-inducible factor-1
IKK	IκBα kinase
IL	interleukin
MAPK	mitogen-activated protein kinase
MEKK-1	MAPK/ERK kinase kinase-1
MMP	matrix metalloproteinase
NF-κB	nuclear factor-κB
NO	nitric oxide
NSCLC	non-small cell lung carcinoma
PDGF	platelet-derived growth factor
PDGFR	platelet-derived growth factor receptor
PI3K	phosphatidylinositol 3′-kinase
Rb	retinoblastoma protein
Src	Src kinase
TGF-α	transforming growth factor-α
TGF-β	transforming growth factor-β
TNF-α	tumor necrosis factor-α
VEGF	vascular endothelial growth factor
VEGFR	vascular endothelial growth factor receptor
XIAP	X-linked inhibitor of apoptosis

Introduction

Angiogenesis and vasculogenesis are distinct physiological processes that govern the formation of new blood vessels through a host of growth factors, integrins, angiopoietins, adhesion and gap junctional proteins, transcription factors, and many other components [1]. During development, vasculogenesis is responsible for the differentiation and growth of blood vessels from embryonic hemangio-blasts into vascular endothelial cells (ECs). As such, ECs, directed by the process of angiogenesis followed by a process of tubulogenesis for the generation of blood-carrying vessels, undergo further differentiation of sprouting or extension along a growth factor gradient [2]. Figure 6.1 shows the most important steps in the process of angiogenesis including: (1) EC activation mediated by hypoxia

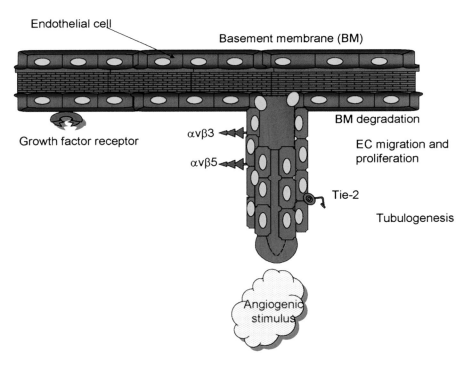

Fig. 6.1 Angiogenesis mechanism in endothelial cells (ECs). In response to an angiogenic stimulus (e.g., tumor, injury, hypoxia, nicotine, and growth factors), vascular ECs become activated, producing and releasing degradative enzymes, and other molecules. Endothelial cells proliferate and migrate toward the angiogenic stimulus. In addition, the basement membrane (BM) is degraded by matrix metalloproteinases (MMPs). The establishment of stable tube structures vasculature (i.e., tubulogenesis) is dependent upon the receptor tyrosine kinase Tie-2-dependent recruitment of stromal cells (e.g., pericytes and smooth muscle cells) by the mature ECs, as well as integrin-mediated cell adhesion to influence growth regulation

and/or cytokine release, (2) degradation of the basement membrane (BM) by matrix metalloproteinases (MMPs), and (3) vascular endothelial growth factor (VEGF)- and integrin-dependent EC migration and proliferation/differentiation. Tubulogenesis involves the maturation of the neovasculature under the influence of angiopoietins and their receptor tyrosine kinase Tie-1 and Tie-2 (Fig. 6.1) [3]. Under pathological conditions, endothelial angiogenesis can contribute to neo-plastic and non-neoplastic transformations including cancer, atherosclerosis, rheumatoid arthritis, and diabetic retinopathy [4]. The trigger for pathophysio-logical angiogenesis is considered to be EC proliferation/dysfunction, by VEGF up-regulation, and possibly through release of stored acetylcholine (ACh) [5]. Considered an autocrine factor in the vascular system, ACh, acting at endothelial nicotinic acetylcholine receptors (AChRs), can modulate vascularization and remodeling [6]. The fact that the acetylcholinesterase inhibitor neostigmine, which increases the endogenous concentration of ACh, accelerates EC migration supports this idea [7].

The most important AChR subtypes involved in the above-mentioned processes are the homomeric α7 AChR and several heteromeric AChRs formed by combinations of α3, α5, β2, and β4 subunits [6], including α3β4-containing AChRs [8]. This chapter will examine the pharmacological evidence for the involvement of non-neuronal AChRs in angiogenesis and how this knowledge may contribute to the development of therapies for angiogenesis-dependent diseases. We will also explore potential mechanisms by which agonists of these receptors, particularly the α7 AChR, mediate endothelial angiogenesis.

AChRs in Angiogenesis

AChRs are members of the Cys-loop superfamily of ligand-gated ion channels mediating rapid synaptic transmission in neurons (reviewed in [9–11]), but may have different functions in non-neuronal, non-excitable cells. Different AChR sub-types are made up of distinct combinations of α and non-α subunits (heteromeric receptors) or of just one type of α subunit (homomeric receptors). In the mamma-lian system, there exist ten α subunits ($\alpha1$-$\alpha10$), four β subunits ($\beta1$-$\beta4$), one δ, and one ε or γ subunit. In this regard, different subunit combinations form specific receptors with distinct physiological and pharmacological properties. AChRs are pentameric glycoproteins, where each subunit has four membrane spanning seg-ments (M1-M4), and the M2 segments form the wall of the ion channel (reviewed in [9–11]). Receptor activation typically causes a net influx of cations (i.e., Ca^{2+}, Na^+, K^+), resulting in membrane depolarization. However, the Ca^{2+} permeability elicited by α7 AChR activation is higher compared with other AChR types [10], supporting the idea that α7 AChRs may trigger Ca^{2+}-mediated intracellular pathways.

Several AChR subunits have been detected in different non-neuronal cells. Table 6.1 lists the expression of various AChR subunits in different human cells

Table 6.1 Nicotinic receptor subunits expressed in non-neuronal tissues in humans

Cell type	Nicotinic subunits	Reference
Vascular endothelial cells	α3, α5, α7, β2-β4	[6, 12–14]
Umbilical vein endothelial cells (HUVECs)	α3-α5, α7, α10, β2-β4	[6, 15]
Bronchial epithelial cells	α1, α3-α7, α9, α10, β1, β2, β4, δ, ε	[8, 13, 16–22]
Airway epithelial cells	α2, α4, α7, α9, α10, β2, β4	[8, 17, 19, 23]
Airway fibroblasts	α1, α5-α7, α9, β1-β3, δ, ε	[16]
Keratinocytes (oral and skin)	α3, α5, α7, α9, α10, β1, β2, β4	[20, 24–26]
Oral epithelium	α3, α5, α7, α9, β2, β4	[11, 27, 28]
Esophageal epithelium	α3, α5, α7, β2	[26]
Brain endothelial cells	α3, α5, α7, β2, β3	[11, 29–31]
Astrocytes	α2-α5, α7, β2, β3	[11, 29]
O2A progenitors	α3-α5, α7, β2	[11, 29]
Glia processes and stroma	α3, α4, β2, β4	[11, 29]
Urothelium	α2-α7, α9, α10	[32]
Lymphocytes	α2-α5, α7, α9, α10, β2, β4	[33–39]
Monocytes	α2, α5-α7, β2	[40]
Macrophages	α1, α7, α10	[41]
Eosinophils	α2, α4, α7	[42]
Bone marrow cells	α4, α7, β2	[43]
Thymocytes	α1, α3, α5, α7, β4	[44–46]
Synoviocytes	α7	[47]
Placenta	α2–α5, α7, α9, α10	[48]

including ECs, epithelial cells, keratinocytes, thymocytes, lymphocytes, microglia, and astrocytes. AChR subunits are also expressed in additional non-neuronal cells (e.g., adipocytes, pancreatic β-cells, etc.) from other species, and probably in humans. Endothelial cells, key players in angiogenesis, express subunits α3, α4, α7, β2, and β4 [6]. Although α7 is the most abundant and the most relevant to the angiogenic process, α3β4-containing AChRs might be also important in ECs from other tissues [8, 12, 22].

Stemming from seminal work by Amparo Villablanca, who provided *in vitro* evidence of nicotine-induced EC proliferation [49], John Cooke and colleagues identified a cholinergic angiogenic pathway [50] mediated by the α7 AChR [6]. In an elegant study they demonstrated that nicotine, a non-selective agonist, induced EC tube formation [6] and migration [51]. Endothelial cell tube formation was also obtained using more selective α7 AChR agonists including choline [15] and dimethoxybenzylidene anabaseine [6], supporting the importance of α7 AChRs in this process. Furthermore, mecamylamine, a nonspecific noncompetitive antagonist [6, 51], as well as selective α7 AChR antagonists such as α-bungarotoxin (α-BTx) [6, 15] and methyllycaconitine [52], significantly reduced the effects elicited by nicotine on ECs from different tissues. Interestingly, antagonists selective for other AChR subtypes such as α-conotoxin MII for α3β2- and α3β3-containing AChRs, α-lobeline for α4β2-containing AChRs, d-tubocurarine for α3β4- and α4β4-containing AChRs, and dihydro-β-erythroidine (DHβE) for α3β2- and α4β2-containing

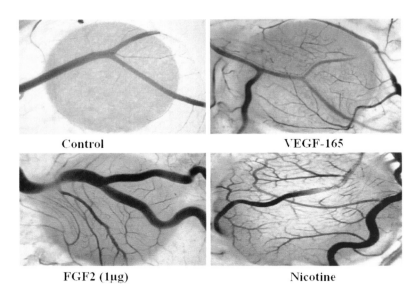

Control   VEGF-165

FGF2 (1µg)   Nicotine

Fig. 6.2 Stimulation of angiogenesis by FGF2, VEGF, and nicotine, using the chick chorioallantoic membrane (CAM) model. The increased neovascularization produced by nicotine is comparable to that of VEGF and FGF2 (bFGF) [54]

AChRs, showed no significant inhibition of capillary network formation [6]. Nevertheless, the inhibition elicited by DHβE on bronchial and airway ECs [8] and on vascular ECs [12] supports the existence of α3β4-containing AChRs.

Further findings suggest a strong connection between growth factor-mediated angiogenic pathways and the cholinergic angiogenic pathway [7]. For instance, using migration assays coupled with micro-array technology, it was recently demonstrated that the EC migration induced by basic fibroblast growth factor (bFGF, FGF2) and VEGF is inhibited by blocking the AChR with α-BTx [51]. In the stimulating of angiogenesis, secreted bFGF directly binds to αvβ3, the integrin receptor that stimulates EC adhesion, to induce spreading and adhesion during angiogenesis (Fig. 6.1). Moreover, bFGF signaling is apparently potentiated when its tyrosine kinase receptor FGFR-1 associates with αvβ3 [53]. Indeed, nicotine-induced angiogenesis *in vivo* was shown to depend upon bFGF, αvβ3, and MAP kinase extracellular signal-regulated kinase 1/2 (ERK1/2) [54]. Figure 6.2 illustrates comparable pro-angiogenic efficacy between bFGF, VEGF, and nicotine, using the chick chorioallantoic membrane (CAM) model. Taking advantage of the fact that αvβ3 is involved in the process of angiogenesis, a humanized monoclonal antibody against αvβ3 (i.e., Vitaxin®) underwent initial drug trials as an angiogenesis inhibitor for malignant tumors, but has yet to show any clinical benefit (reviewed in [55]).

VEGF and bFGF are both included in a group of at least 20 known angiogenic factors, but VEGF, one of the most potent factors, is a key target in therapeutic approaches to pathological angiogenesis. However, the molecular mechanisms of angiogenesis, and arguably, the cholinergic angiogenic pathway, are governed by VEGF-dependent and -independent signaling. Angiogenesis is important for a number of normal processes, including wound-healing and reproduction. Angiogenesis inhibitors generally have mild side effects, but underlying conditions, such as heart disease or hypertension may be exacerbated in the presence of angiogenesis inhibition therapy. These inhibitors fall into two general categories: those that act indirectly to either clear circulating angiogenic growth factors or prevent growth factor signaling and those that act directly on the endothelium and ECs [56]. Nicotine, a non-selective cholinergic agonist, has been shown to induce angiogenesis, and mecamylamine, a nonspecific cholinergic antagonist, has been shown to inhibit EC proliferation and ultimately blood vessel formation; as a result, there has been much interest in defining the molecular mechanisms of these actions, with the ultimate purpose of developing direct-acting angiogenesis promoters and inhibitors. In this regard, the vascular EC ChRs, which expresses a number of AChRs, is an attractive therapeutic target.

AChRs and Tumor Angiogenesis

Typically tumors use simple diffusion (up to 1–2 mm) to obtain oxygen and nutrients, but further growth requires communication with vascular ECs and the formation of new blood vessels. Hence, malignancy is dependent upon an "angiogenic switch", a consequence of an imbalance of pro- and anti-angiogenic factors influenced by hypoxia, low pH, hypoglycemia, and inflammatory cytokines [57, 58]. Tumor cells secrete a host of angiogenic factors, which can no longer be balanced by endogenous angiogenic inhibitors. In an apparently coordinated fashion, ECs migrate toward such an angiogenic stimulus, at which point they proliferate and initiate vessel formation (Fig. 6.1). However, tumor ECs, which exist in a hypoxic micro environment, proliferate 20–2,000 times faster than normal ECs [3]. With these processes in mind, several angiogenesis inhibitors were developed as cancer therapeutic agents, with a few having obtained FDA approval. A number of laboratories have focused efforts on the molecular mechanism of nicotine-induced tumor cell proliferation and angiogenesis, and have identified differential expression of AChRs in cancer cells [59, 60]. In particular, $\alpha7$ AChR is the primary receptor type that mediates the proliferative effect of nicotine in the tumor cell. Nevertheless, $\alpha3\beta4$-containing AChRs might also be important in this process [8]. The non-selective AChR agonist nicotine, the main active component in cigarettes, enhances the proliferation of tumor cells, including small cell and non-small cell lung carcinoma (NSCLC), and also pancreatic, colon, and bladder cancer cells [61–64]. Similarly, in the CAM tumor implant model, nicotine accelerated the growth of breast, lung, and colon cancer cells, and also induced neovascularization in a dose-dependent fashion [54]. Indeed, nicotine has been shown to increase levels of different growth factors including VEGF, brain-derived neurotrophic factor (BDNF), epidermal growth factor (EGF), hepatocyte growth factor (HGF), platelet-derived growth factor (PDGF), and transforming growth factor-α (TGF-α) and β (TGF-β), as well as their respective receptors [18], which influence both tumor and

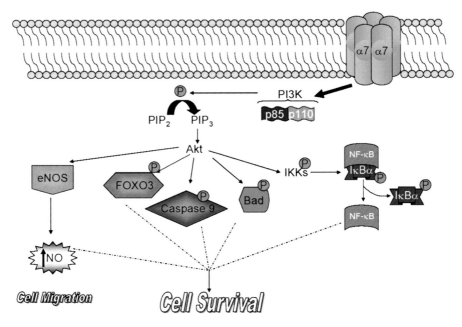

Fig. 6.3 Cell survival pathways. The PI3K/Akt pathway is a prototypic cell survival pathway found in normal and tumor cells. Once activated, Akt can signal through a diverse array of substrates. Akt-mediated phosphorylation (P) inactivates pro-apoptotic caspase 9, and the Bcl2 family member Bad, whereas it activates, through IκBα kinase (IKK) phosphorylation, the pro-survival transcription factors FOXO3 (forkhead box O3) and NF-κB. Akt also activates endothelial nitric oxide synthase (eNOS) for the production of nitric oxide (NO) and subsequent cell migration

normal ECs. Heeschen and colleagues used a mouse lung tumor model to show that a significant decrease in tumor size resulted after treatment with the nonspecific AChR antagonist mecamylamine [6]. Moreover, this decrease in tumor burden corresponded to lower capillary density and decreased systemic VEGF levels.

The mechanism of nicotine-induced tumor cell proliferation and associated angiogenesis is an active area of research. However, tumor and normal cells may exploit different mechanisms to transduce signals from activated AChRs. On the other hand, there may be crosstalk between signal transduction pathways [65]. For example, Heeschen and co-workers, using an *in vitro* angiogenesis assay, were able to show that inhibition of MAP kinase (ERK), p38 MAPK, and phosphoinositide 3′-kinase (PI3K), completely prevented α7 AChR-induced EC tube network formation [6]. Interestingly, the reduction in tube formation was slightly greater with the inhibition of PI3K than that with the inhibition of other kinases. A key mediator in cell survival, PI3K, in response to external signals, governs changes in a broad range of cellular functions. Proteins phosphorylated by its downstream effector molecule Akt (also known as protein kinase B) promote cell survival through a variety of substrates including caspases, transcription factors, and enzymes (Fig. 6.3). Indeed, the PI3K/Akt pathway is considered an important target in cancer chemotherapy [66]. Nevertheless, in normal bronchial and airway ECs, specific α7 antagonists–including α-BTx, methyllycaconitine and the nonspecific AChR antagonist, mecamylamine did not attenuate nicotine-induced Akt phosphorylation. This suggested the existence of other intracellular pathways in these normal cells [8]. One critical aspect of the PI3K/Akt pathway is that the phosphorylation of IκBα kinase (IKK) by Akt leads to the activation of nuclear factor-κB (NF-κB) (Fig. 6.3), which is primarily involved in the transcription of genes to block apoptosis and hence contribute to cell survival. NF-κB along with hypoxia-inducible factor-1α (HIF-1α), and coordinated by the hypoxic micro environment, are considered master regulators of tumor-associated angiogenesis [67]. NF-κB activation can also lead to inhibition of tumor angiogenesis, but the mechanism has yet to be determined [68]. Nicotine itself can stimulate HIF-1α protein accumulation and VEGF expression in human NSCLC, contributing to increased tumor angiogenesis and invasion by activating various downstream AChR-mediated signaling pathways, including Ca^{2+}/calmodulin, protein kinase C (PKC), PI3K/Akt MAPK/ERK1/2, mTor (mammalian target of rapamycin), and Src kinase (Src) [69]. Src, a proto-oncogenic tyrosine kinase similar to the Rous sarcoma virus, has been gaining relevance to nicotine-mediated tumor angiogenesis. Src promotes the expression of VEGF in tumor cells as well as maintains vascular barrier function in ECs. Furthermore, Src-dependent hypoxia-mediated VEGF expression in human tumor cells is regulated by a transcriptional complex of HIF-1α, STAT3 (signal transducer and activator of transcription 3), CREB binding protein (CBP)/p300, and by Ref1/APE (redox-factor 1/apurinic: apyrimidinic endonuclease) [70].

With respect to tumor cell proliferation in human NSCLC, nicotine, through the PI3K/Akt survival pathway, up-regulates proteins such as survivin and X-linked inhibitor of apoptosis (XIAP), blocking chemotherapeutic drug-induced apoptosis [71]. In addition to these anti-apoptotic gene products, nicotine may also mediate multi-site phosphorylation of the Bcl2 (B-cell lymphoma 2) family members Bad and Bax in a mechanism involving the activation of ERK1/2, PI3K/Akt, and protein kinase A [72, 73]. Intriguingly, there is a negative association between Bcl2 and microvessel counts in patients with NSCLC that suggests a regulatory role of Bcl2 on VEGF expression [74]. Moreover, activation of α7 AChRs in lung mesothelioma cells results in Ca^{2+} influx and subsequent activation of MAPK/ERK kinase kinase-1 (MEKK-1), ERK1/2, and p90RSK (p90 ribosomal S6 kinase), a downstream target of phosphorylated ERK1/2 and possible component of VEGF-mediated gene transcription [75, 76].

The mechanism of AChR-mediated cellular proliferation was given greater clarification when Dasgupta and colleagues, using siRNA and immunoblotting techniques, nicely showed that nicotine-mediated activation of α7 AChRs triggered NSCLC proliferation by promoting cell cycle entry [17]. Such activation led to the recruitment to the receptor of the scaffolding protein β-arrestin, facilitating the activation of Src and subsequently leading to binding of Raf-1 kinase (i.e., a Ser/Thr-specific kinase) to and inactivation of the retinoblastoma protein (Rb) [17]. The inactivation of Rb is seen in a number of cancers, and through association with cyclins D and E, leads to the activation of E2F-regulated proliferative promoters and subsequent entry into cell S-phase [77].

Given the intense research and promising data on the signaling pathways, the development of specific, potent AChR ligands will facilitate the design of new angiogenesis inhibitors for cancer therapy.

AChRs and Ischemic Revascularization

Cellular repair after ischemic injury–such as stroke, myocardial infarction, diabetic retinopathy, and age-related macular degeneration (AMD)–relies on the regeneration of damaged and depleted vasculature to improve blood flow and subsequent distribution of oxygen and nutrients, and the removal of waste metabolites. During ischemic-dependent angiogenesis, there is an up-regulation of endothelial nitric oxide synthase (eNOS) and subsequent generation of nitric oxide (NO), a known regulator of EC migration [78] (Fig. 6.3). Nicotine can stimulate NO activity, which is apparently dependent on interactions between NO and either NADPH or oxygen radicals [79]. In addition, in the murine model of hind-limb ischemia, in which the superficial and deep femoral arteries are unilaterally ligated, Heeschen and colleagues administered mecamylamine over a 3 week period [6]. Although no change was seen in capillary density in the control hind limb, there was a significant reduction in the angiogenic response in the ischemic hind limb. They were also able to confirm that the α7 AChR subtype was up-regulated in the ischemic tissue and co-localized with ECs. Similarly, after coronary artery ligation to experimentally induce myocardial infarction in the rat, it was demonstrated that the α7 AChR subtype was up-regulated in the damaged region and was associated with significant increases in capillary density [15]. Furthermore, after 3 weeks of post-ligation mecamylamine treatment, capillary density was decreased in the infarct region. In a complementary study in support of these findings, using laser Doppler perfusion imaging, nicotine was shown to improve blood flow by nearly 75 % following endothelial progenitor cell transplants in an animal model of ischemic disease [80]. Although the molecular mechanism of AChR-mediated angiogenesis was not the focus of either of these studies, it was suggested that VEGF induction in ECs may be a contributing factor, but it is distinct from the cholinergic angiogenesis pathway [6, 80]. As ischemia invariably results in tissue hypoxia, the induction of VEGF is, in part, regulated by HIF-1α and NO [81].

Ocular blood flow is regulated by NO derived in part from the endothelium, and thus hypoperfusion of the eye can result in ischemic events and blindness. Pathological angiogenesis in the retina is characterized by the overgrowth of leaky, disorganized and non-functional blood vessels [82]. For example, in AMD there is proliferation of abnormal new vessels from the choroid vascular network that lies beneath the retinal pigment epithelium; this is referred to as choroidal neovascularization. After receiving nicotine in their water for 4 weeks, mice showed greater and more severe choroidal neovascularization than control animals, with the effect blocked by hexamethonium, a nonspecific AChR antagonist [83]. Similarly, but more intriguing, mecamylamine was able to block choroidal neovascularization *in vivo* with or without orally administered nicotine, suggesting that an endogenous activation of AChR in the promotion of this abnormal vessel outgrowth exists in animals not treated with nicotine [84]. In both studies it is strongly suggested that VEGF-dependent processes were involved in the AChR signaling, and that VEGF may stimulate retinal neovascularization. AChR antagonism with mecamylamine suppressed VEGF-induced tube formation by retinal and choroidal ECs [84], whereas nicotine reversed VEGF-induced suppression of MMP-2 activity in choroidal vascular smooth muscle cells [83]. Despite encouraging outcomes, the current therapy for AMD is based on chronic intravitreal administration of VEGF inhibitors, an obvious treatment burden. Thus, investigations into alternative, safer therapy with AChR antagonists may prove beneficial. In this regard, mecamylamine (Inversine®) has been used as an orally antihypertensive agent for decades [85] and might be a plausible therapy.

AChRs and Inflammatory Angiogenesis

Agonists acting at α7 AChRs may prove beneficial in the treatment of epilepsy as anticonvulsants [86], but may also have some use in therapeutic angiogenesis. After seizures, there is breakdown of the blood brain barrier, a network of tight junctions between brain capillary ECs, as well as aberrant angiogenesis, and inflammation [87, 88]. Overall, inflammatory angiogenesis, in which blood vessels enter a lesion caused by infiltrating immune cells, is the hallmark of a number of pathological conditions including atherosclerosis, diabetes, psoriasis, asthma, rheumatoid arthritis, and neurodegenerative diseases, as well as immune cell-associated tumor growth (reviewed in [3]). A pathway has been described in which α7 AChR activation of macrophages [41, 89, 90] can significantly inhibit the release of pro-inflammatory cytokines such as tumor necrosis factor-α (TNF-α), and interleukin-1, -6, and -8 (IL-1, -6, and -8, respectively) [91]. This mechanism, termed the 'cholinergic anti-inflammatory pathway', may also involve mononuclear lymphocytes and dendritic cells [90], microglia [92], eosinophils [42], and mast cells [93].

The respiratory and cardiovascular system, and possibly the central nervous system, are influenced by this 'cholinergic anti-inflammatory pathway' as well, which presents major implications for immunotherapeutic approaches. For example, asthma, a classic case of airway inflammation, is characterized by massive infiltration of eosinophils, mast cells, and macrophages, as well as by the presence of hyper-vascularization [94]. Angiogenesis and inflammation are co-dependent processes, in the sense that inflammatory mediators promote directly and/or indirectly angiogenesis, and in that angiogenesis contributes to inflammatory pathology [95]. For example, eosinophils, by releasing pro-angiogenic mediators such as VEGF, promote EC proliferation, potentially inducing the neovascularization seen in asthma. Interestingly, human blood eosinophils express α3, α4, and α7 AChR subunits (Table 6.1), and their migration (a key event in the pathogenesis of asthma) can be inhibited by 1, 1-dimethyl-4-phenyl-piperazinium (DMPP), a non-selective AChR agonist [42]. The exact mechanism of the anti-inflammatory effect mediated by DMPP in eosinophils is not completely clear, but involves the reduction of platelet-activating factor (PAF)-induced LTC4 production, eotaxin, and 5-oxo-6,8,11,14-eicosatetranoic acid-induced eosinophil migration [42].

In alveolar macrophages, DMPP, which activates α4- and α7-containing AChRs, inhibits TNF-α (a pro-inflammatory/pro-angiogenic cytokine) production through activation of PI3K and PLC [96]. Although both cells are of hematopoietic origin, macrophages and eosinophils respond differently to ACh, which is anti-inflammatory for the former and both pro- and anti-inflammatory for the latter [97]. Indeed, it has been recently demonstrated that PI3K activity is essential for hematopoetic progenitor survival and that high activity of its signaling partner Akt promotes monocyte/macrophage development, while reduced Akt activity induces eosinophil differentiation [98]. Thus, it is plausible that the anti-inflammatory effect of DMPP in eosinophils is also produced through the PI3K pathway.

TNF-α is also a known potent activator of NF-κB. *In vitro* data suggest that the decreased activation of NF-κB, by relieving negative regulation, can reduce EC tube formation [99]. Furthermore, TNF-α *in vitro* induces a "tip-cell" phenotype in vascular ECs, thus priming them for angiogenic sprouting in an NF-κB-dependent fashion [100]. Hence, an inhibition of angiogenesis could potentially lessen chronic inflammation by attenuating the recruitment of immune cells. Depending on the concentration and duration, TNF-α can induce EC dysfunction by decreasing NO and increasing endothelial permeability [101], contributing to the progression of atherosclerosis. In atherosclerotic cardiovascular disease, in which the accumulation of macrophages is a fundamental event, there is a connection between pathological angiogenesis and the growth of atherosclerotic plaques [102]. However, scant information on the link between α7 AChR activation and macrophage accumulation/EC migration in the atheroma exists [7]. Nevertheless, in a rabbit model of balloon-injured aortas, intramuscularly administered nicotine significantly increased the expression of VEGF, promoting intramyocardial angiogenesis [103]. Also, in the Apolipoprotein E-deficient mouse model of atherosclerosis, nicotine was shown to increase the size of atherosclerotic lesions, in association with an increase in plaque vascularization [50]. This effect was reversed by the selective cyclooxygenase-2 (COX-2) inhibitor rofecoxib. COX-2 is present on nascent ECs, and its selective inhibition can be anti-angiogenic through the prevention of new adhesion formation [104]. However, nicotine can inhibit the production of pro-inflammatory cytokines from monocytes/macrophages through a mechanism that includes α7 AChR-dependent up-regulation of COX-2 [105]. Hence, this is a protective mechanism suitable for therapeutic exploitation.

Interestingly, nicotine-induced neuroprotection may be mediated by an interaction between α7 AChRs and TNF-α [106]. It seems reasonable to expect a similar interaction between non-neuronal α7 AChRs and TNF-α. For example, nicotine has been shown to inhibit the release of TNF-α from microglia, the phagocytic immune cells from the brain [85, 107]. Activated microglia are implicated in the pathogenesis of a number of neurodegenerative diseases including Alzheimer's disease [108], multiple sclerosis, amyotrophic lateral sclerosis, and Parkinson's disease. However, there is evidence suggesting that activated microglia may also serve a protective role [109].

Microglia, which share a common myeloid lineage with macrophages, do release and respond to the angiogenic factors VEGF, bFGF, and PDGF, but linkage between these cells and neuroinflammatory angiogenesis is sparse. However, it has been suggested that microglia, like macrophages, contribute to blood vessel formation by secreting MMPs for the degradation of the vascular basement membrane [110], and are required for the maintenance of immature blood vessels and the growth of new ones [111]. Although considered "non-excitable", these immune cells do express the same voltage-gated potassium channel, i.e., Kir1.3 [112], as in human ECs (important for VEGF-induced proliferation) [113]; this contributes to an electrical behavior in microglia that may reveal an even greater role in the cholinergic anti-inflammatory/angiogenesis pathway.

AChR and Wound-healing Angiogenesis

Angiogenesis and increased vascular permeability are important features of the wound-healing process. As with angiogenesis, wound-healing is a tightly controlled multifactorial process that involves re-epithelialization, macrophage accumulation, fibroblast infiltration, matrix formation, and revascularization [114]. Controversy exists over whether nicotine promotes or delays wound-healing, however, the concentration or dose may be the determinant of the nature of the effect. Epithelial cells lining the skin and oral mucosa, and the lung, express $\alpha3$, $\alpha4$, $\alpha5$, $\alpha7$, $\alpha9$, $\alpha10$, $\beta1$, $\beta2$, and $\beta4$ subunits [8, 115, 116] (Table 6.1). The therapeutic potential in modulating these subunits can be seen with both neoplastic and non-neoplastic diseases such as airway epithelial injury, diabetic ulcers, bone fractures, and gastrointestinal bleeding. For example, the use of an excision wound technique revealed that low-dose nicotine increased the rate of wound closure and produce greater vascularity at the wound site in diabetic [117] and non-diabetic mice [118]. Wound closure is a re-epithelialization that requires the migration and interaction of keratinocytes and fibroblasts, as well as the release of adhesion molecules and growth factors [119]. Recently, in a well-designed study, an early key regulator of this step has been identified *in vivo* and *in vitro* as the $\alpha9$ AChR by up-regulating the expression of adhesion molecules [24, 120]. The effects were not observed in $\alpha9$ knockout animals and were reversed when siRNA against the $\alpha9$ subtype was employed in keratinocyte cultures.

Bone, like skin, is a highly vascularized tissue and is thus dependent upon a close association between angiogenesis and osteogenesis. The expression profile of AChRs on bone ECs, or vascular ECs in bone, is currently unknown. In a rabbit model of bone healing and regeneration, chronic exposure to nicotine increased the density of microvessels in the vascular endothelium surrounding the wound site [121]. Again, although not investigated, it was suggested in this study that VEGF is mediating the nicotine angiogenic response. In light of recent data suggesting that in response to norepinephrine (an endogenous vasoconstrictor) nicotine can alter bone function by enhancing constriction of blood vessels that supply the tibia [122], additional investigations into nicotine-mediated bone vascularization are needed.

Conclusions

There is substantial interest in non-neuronal AChRs for their roles in angiogenesis, and as potential drug targets for neoplastic and vascularization-related disease states. Current anti-angiogenic drug therapies under study or in clinical trials have met with variable success. Angiogenesis inhibitors can act indirectly by clearing circulating angiogenic growth factors or by suppressing their signaling pathways and directly by acting on the endothelium and ECs, to affect regulatory pathways [123]. Drugs that target ECs such as endostatin, 2-methoxyestradiol, thrombospondin, and IFN-α, can become ineffective due to acquired resistance. Tumor-associated ECs can have cytogenetic abnormalities, as well as express multidrug resistance P-glycoproteins, mediated by high levels of VEGF and other EC survival factors, resulting in an up-regulation of anti-apoptotic signals and transcription factors such as HIF-1α [124–126]. Specific VEGF inhibitors such as bevacizumab can exhibit a wide array of adverse effects in humans when used as anti-cancer treatments, including the risk of thrombosis.

In this chapter, we have discussed the experimental evidence indicating that nicotine, a non-selective AChR agonist, produces a clear pro-angiogenic stimulus in a number of *in vitro* and *in vivo* models. However, this compound is not suitable for therapeutic use due to potential adverse effects. In addition, non-selective antagonists such as mecamylamine are undoubtedly anti-angiogenic in nature. For example, based on promising rodent data showing that lower-dose mecamylamine (50 mg/kg/day infusion or 1 % topical administration) blocks choroidal neovascularization [84], phase II clinical trials using ophthalmic solutions of mecamylamine (ATG003) (CoMenits, South San Francisco, CA, USA) have been completed for the treatment of AMD.

The α7 AChR subtype has been successfully linked to the angiogenic process as it relates to tumor vascularity, inflammation, ischemia, and wound repair, and is becoming an attractive target for more selective and potent agonists and antagonists for angiogenic therapeutics. For example, JN403, a potent, partial agonist of the α7 AChR, has promising properties for the treatment of several neurological problems, without adverse consequences [86]. In this regard, preliminary results from our laboratory ascribe a pro-angiogenic activity elicited to JN403, with potential use in wound-healing. In addition, clinical trials using benzylidene-anabaseine analogues have demonstrated beneficial activity for the treatment of schizophrenia and Alzheimer's disease, with virtually no side effects [127]. Thus, these and other new untested compounds [13] could also be examined for angiogenic therapy. The development of such drugs could be beneficial for angiogenesis-associated diseases.

References

1. Risau W (1997) Mechanism of angiogenesis. Nature 386:671–6741
2. Coultas L, Chawengsaksophak K, Rossant J (2005) Endothelial cells and VEGF in vascular development. Nature 438:937–945

3. Griffioen AW, Molema G (2000) Angiogenesis: potentials for pharmacologic intervention in the treatment of cancer, cardiovascular diseases, and chronic inflammation. Pharmacol Rev 52:237–268

4. Folkman J (2006) Angiogenesis. Annu Rev Med 57:1–18

5. Parnavelas JG, Kelly W, Burnstock G (1985) Ultrastructural localization of choline aceytltransferase in vascular endothelial cells in rat brain. Nature 316:724–725

6. Heeschen C, Weis M, Aicher A, Dimmeler S, Cooke JP (2002) A novel angiogenic pathway mediated by non-neuronal nicotinic acetylcholine receptors. J Clin Invest 110:527–536

7. Cooke JP (2007) Angiogenesis and the role of the endothelial nicotinic acetylcholine receptor. Life Sci 80:2347–2351

8. West KA, Brognard J, Clark AS, Linnoila IR, Yang X, Swain SM, Harris C, Belinsky S, Dennis PA (2003) Rapid Akt activation by nicotine and a tobacco carcinogen modulates the phenotype of normal human airway epithelial cells. J Clin Invest 111:81–90

9. Arias HR (2006) Ligand-gated ion channel receptor superfamilies. In: Arias HR (ed) Biological and biophysical aspects of ligand-gated ion channel receptor superfamilies, Research Signpost, Kerala, Chapter 1, pp 1–25

10. Arias HR, Bhumireddy P, Bouzat C (2006) Molecular mechanisms and binding site locations for noncompetitive antagonists of nicotinic acetylcholine receptors. Int J Biochem Cell Biol 38:1254–1276

11. Gotti C, Clementi F (2004) Neuronal nicotinic receptors: from structure to pathology. Prog Neurobiol 74:363–396

12. Macklin KD, Maus AD, Pereira EF, Albuquerque EX, Conti-Fine BM (1998) Human vascular endothelial cells express functional nicotinic acetylcholine receptors. J Pharmacol Exp Ther 287:435–439

13. Wang Y, Pereira EF, Maus AD, Ostlie NS, Navaneetham D, Lei S, Albuquerque EX, Conti-Fine BM (2001) Human bronchial epithelial and endothelial cells express α7 nicotine acetylcholine receptors. Mol Pharmacol 60:1201–1209

14. Saeed RW, Varma S, Peng-Nemeroff T, Sherry B, Balakhaneh D, Huston J, Tracey KJ, Al-Abed Y, Metz CN (2005) Cholinergic stimulation blocks endothelial cell activation and leukocyte recruitment during inflammation. J Exp Med 201:1113–1123

15. Li XW, Wang H (2006) Non-neuronal nicotinic α7 receptor, a new endothelial target for revascularization. Life Sci 78:1863–1870

16. Carlisle DL, Hopkins TM, Faither-Davis A, Silhanek MJ, Luketich JD, Christie NA, Siegfried JM (2004) Nicotine signals through muscle-type and neuronal nicotinic acetylcholine receptors in both human bronchial epithelial cells and airway fibroblasts. Respir Res 5:27

17. Dasgupta P, Rastogi S, Pillai S, Ordonez-Ercan D, Morris M, Haura E, Chellappan S (2006) Nicotine induces cell proliferation by β-arrestin-mediated activation of Src and Rb-Raf-1 pathways. J Clin Invest 116:2208–2217

18. Conti-Fine BM, Navaneetham D, Lei S, Maus AD (2000) Neuronal nicotinic receptors in non-neuronal cells: new mediators of tobacco toxicity? Eur J Pharmacol 393:279–294

19. Lam DC, Girard L, Ramirez R, Chau W, Suen W, Sheridan S, Tin VPC, Chung L, Wong MP, Shay JW, Gazdar AF, Lam W, Minna JD (2007) Expression of nicotinic acetylcholine receptor subunit genes in non-small-cell lung cancer reveals differences between smokers and nonsmokers. Cancer Res 67:4638–4647

20. Zia S, Ndoye A, Lee TX, Webber RJ, Grando SA (2000) Receptor-mediated inhibition of keratinocyte migration by nicotine involves modulations of calcium influx and intercellular concentration. J Pharmacol Exp Ther 293:973–981

21. Tsurutani J, Castillo SS, Brognard J, Granville CA, Zhang C, Gills JJ, Sayyah J, Dennis PA (2005) Tobacco components stimulate Akt-dependent proliferation and NFκB-dependent survival in lung cancer cells. Carcinogenesis 26:1181–1195

22. Maus AD, Pereira EF, Karachunski PI, Horton RM, Navaneetham D, Macklin K, Cortes WS, Albuquerque EX, Conti-Fine BM (1998) Human and rodent bronchial epithelial cells express functional nicotinic acetylcholine receptors. Mol Pharmacol 54:779–788

23. Plummer HK III, Dhar M, Schuller HM (2005) Expression of the α7 nicotinic acetylcholine receptor in human lung cells. Respir Res 6:29–37

24. Chernyavsky AI, Arredondo J, Marubio LM, Grando SA (2004) Differential regulation of keratinocyte chemokinesis and chemotaxis through distinct nicotinic receptor subtypes. J Cell Sci 117:5665–5679

25. Arredondo J, Chernyavsky AI, Jolkovsky DL, Pinkerton KE, Grando SA (2007) Receptor-mediated tobacco toxicity: alterations of the NF-κB expression and activity downstream of α7 nicotinic receptor in oral keratinocytes. Life Sci 80:2191–2194

26. Nguyen VT, Hall LL, Gallacher G, Ndoye A, Jolkovsky DL, Webber RJ, Buchli R, Grando SA (2000) Choline acetyltransferase, acetylcholinesterase, and nicotinic acetylcholine receptors of human gingival and esophageal epithelia. J Dent Res 79:939–949

27. Arredondo J, Nguyen VT, Chernyavsky AI, Bercovich D, Orr-Urtreger A, Kummer W, Lips K, Vetter DE, Grando SA (2002) Central role of α7 nicotinic receptor in differentiation of the stratified squamous epithelium. J Cell Biol 159:325–336

28. Arredondo J, Chernyavsky AI, Marubio LM, Beaudet AL, Jolkovsky DL, Pinkerton KE, Grando SA (2005) Receptor-mediated tobacco toxicity: regulation of gene expression through α3β2 nicotinic receptor in oral epithelial cells. Am J Pathol 166:597–613

29. Sharma G, Vijayaraghavan S (2001) Nicotinic cholinergic signaling in hippocampal astrocytes involves calcium-induced calcium release from intracellular stores. Proc Natl Acad Sci USA 98:4148–4153

30. Abbruscato TJ, Lopez SP, Roder K, Paulson JR (2004) Regulation of blood–brain barrier Na, K,2Cl-cotransporter through phosphorylation during in vitro stroke conditions and nicotine exposure. J Pharmacol Exp Ther 310:459–468

31. Hawkins BT, Egleton RD, Davis TP (2005) Modulation of cerebral microvascular permeability by endothelial nicotinic acetylcholine receptors. Am J Physiol Heart Circ Physiol 289:H212–H219

32. Bschleipfer T, Schukowski K, Weidner W, Grando SA, Schwantes U, Kummer W, Lips KS (2007) Expression and distribution of cholinergic receptors in the human urothelium. Life Sci 80:2303–2307

33. Peng H, Feris RL, Matthews T, Hiel H, Lopez-Albaitero A, Lustig LR (2004) Characterization of the human nicotinic acetylcholine receptor subunit alpha (α) 9 (CHRNA9) and alpha (α) 10 (CHRNA10) in lymphocytes. Life Sci 76:263–280

34. Kawashima K, Fujii T (2003) The lymphocytic cholinergic system and its contribution to the regulatin of immune activity. Life Sci 74:675–696

35. Razani-Boroujerdi S, Boyd RT, Dávila-García MI, Nandi JS, Mishra NC, Singh SP, Pena-Philippides JC, Langley R, Sopori ML (2007) T Cells express α7-nicotinic acetylcholine receptor subunits that require a functional TCR and leukocyte-specific protein tyrosine kinase for nicotine-induced Ca²⁺ response. J Immunol 179:2889–2898

36. Skok MV, Grailhe R, Agenes F, Changeux J-P (2007) The role of nicotinic receptors in B-lymphocyte development and activation. Life Sci 80:2334–2336

37. De Rosa MJ, Esandi MC, Garelli A, Rayes D, Bouzat C (2005) Relationship between α7 nAChR and apoptosis in human lymphocytes. J Neuroimmunol 160:154–161

38. Battaglioli E, Gotti C, Terzano S, Flora A, Clementi F, Fornasari D (1998) Expression and transcriptional regulation of the human α3 neuronal nicotinic receptor subunit in T lymphocyte cell lines. J Neurochem 71:1261–1270

39. Hiemke C, Stolp M, Reuss S, Wevers A, Reinhardt S, Maelicke A, Schlegel S, Schröder H (1996) Expression of α subunit genes of nicotinic acetylcholine receptors in human lymphocytes. Neurosci Lett 214:171–174

40. Sato KZ, Fujii T, Watanabe Y, Yamada S, Ando T, Fujimoto K, Kawashima K (1999) Diversity of mRNA expression for muscarinic acetylcholine receptor subtypes and neuronal nicotinic acetylcholine receptor subunits in human mononuclear leukocytes and leukemic cell lines. Neurosci Lett 266:17–20

41. Wang H, Yu M, Ochani M, Amella CA, Tanovic M, Susarla S, Li JH, Wang H, Yang H, Ulloa L, Al-Abed Y, Czura CJ, Tracey KJ (2003) Nicotinic acetylcholine receptor α7 subunit is an essential regulator of inflammation. Nature 421:384–388

42. Blanchet MR, Langlois A, Israel-Assayag E, Beaulieu MJ, Ferland C, Laviolette M, Cormier Y (2007) Modulation of eosinophil activation *in vitro* by a nicotinic receptor agonist. J Leukoc Biol 81:1245–1251

43. Koval LM, Zverkova AS, Grailhe R, Utkin YN, Tsetlin VI, Komisarenko SV, Skok MV (2008) Nicotinic acetylcholine receptors α4β2 and α7 regulate myelo- and erythropoiesis within the bone marrow. Int J Biochem Cell Biol 40:980–990

44. Mihovilovic M, Denning S, Mai Y, Fisher CM, Whichard LP, Patel DD, Roses AD (1998) Thymocytes and cultured thymic epithelial cells express transcripts encoding α-3, α-5, and β-4 subunits of neuronal nicotinic acetylcholine receptors. Preferential transcription of the α-3 and β-4 genes by immature CD⁺48⁺ thymocytes and evidence for response to nicotine in thymocytes. Ann N Y Acad Sci 841:388–392

45. Flora A, Schulz R, Benfante R, Battaglioli E, Terzano S, Clementi F, Fornasari D (2000) Neuronal and extraneuronal expression and regulation of the human α5 nicotinic receptor subunit gene. J Neurochem 75:18–27

46. Navaneetham D Jr, Penn A Jr, Howard J Jr, Conti-Fine BM (1997) Expression of the α7 subunit of the nicotinic acetylcholine receptor in normal and myasthenic human thymuses. Cell Mol Biol 43:433–442

47. Waldburger J-MM, Boyle DLL, Pavlov VAA, Tracey KJJ, Firestein GSS (2008) Acetylcholine regulation of synoviocyte cytokine expression by the α7 nicotinic receptor. Arthr Rheum 58:3439–3449

48. Lips KS, Brüggmannb D, Pfeila U, Vollerthuna R, Grandoc SA, Kummera W (2005) Nicotinic acetylcholine receptors in rat and human placenta. Placenta 26:735–746

49. Villablanca A (1998) Nicotine stimulates DNA synthesis and proliferation in vascular endothelial cells in vitro. J Appl Physiol 84:2089–2098

50. Heeschen C, Jang JJ, Weis M, Pathak A, Kaji S, Hu RS, Tsao PS, Johnson FL, Cooke JP (2001) Nicotine stimulates angiogenesis and promotes tumor growth and atherosclerosis. Nat Med 7:833–839

51. Ng MKC, Wu J, Chang E, Wang BY, Katzenberg-Clark R, Ishii-Watabe A, Cooke JP (2007) A central role for nicotinic cholinergic regulation of growth factor-induced endothelial cell migration. Arterioscler Thromb Vasc Biol 27:106–112

52. Beckel JM, Kanai A, Lee SJ, de Groat WC, Birder LA (2006) Expression of functional nicotinic acetylcholine receptors in rat urinary bladder epithelial cells. Am J Physiol Renal Physiol 290:F103–F110

53. Murakami M, Elfenbein A, Simons M (2008) Non-canonical fibroblast growth factor signaling in angiogenesis. Cardiovasc Res 78:223–231

54. Mousa S, Mousa SA (2006) Cellular and molecular mechanisms of nicotine's pro-angiogenesis activity and its potential impact on cancer. J Cell Biochem 97:1370–1378

55. Rüegg C, Hasmim M, Lejeune FJ, Alghisi GC (2006) Antiangiogenesis peptides and proteins: from experimental tools to clinical drugs. Biochim Biophys Acta 1765:155–177

56. Tabruyn SP, Griffioen AW (2007) Molecular pathways of angiogenesis inhibition. Biochem Biophys Res Commun 355:1–5

57. Carmeliet P, Jain RK (2000) Angiogenesis in cancer and other diseases. Nature 407:249–257

58. Pandya NM, Dhalla NS, Santani DD (2006) Angiogenesis – a new target for future therapy. Vascul Pharmacol 44:265–274

59. Egleton RD, Brown KC, Dasgupta P (2008) Nicotinic acetylcholine receptors in cancer: multiple roles in proliferation and inhibition of apoptosis. Trends Pharmacol Sci 29:151–158

60. Chini B, Clementi F, Hukovic N, Sher E (1992) Neuronal-type α-bungarotoxin receptors and the α5-nicotinic receptor subunit gene are expressed in neuronal and human cell lines. Proc Natl Acad Sci USA 89:1572–1576

61. Wong HP, Yu L, Lam EK, Tai EK, Wu WK, Cho CH (2007) Nicotine promotes cell proliferation via α7-nicotinic receptor and catecholamine-synthesizing enzymes-mediated pathway in human colon adenocarcinoma HT-29 cells. Toxicol Appl Pharmacol 221:261–2671

62. Shin VY, Wu WK, Ye YN, So WH, Koo MW, Liu ES, Luo JC, Cho CH (2004) Nicotine promotes gastric tumor growth and neovascularization by activating extracellular signal-related kinase and cyclooxygenase-2. Carcinogenesis 25:2487–2495

63. Ye YN, Liu ES, Shin VY, Wu WK, Luo JC, Cho CH (2004) Nicotine promoted colon cancer growth via epidermal growth factor receptor, c-Src, and 5-lipoxygenase-mediated signal pathway. J Pharmacol Exp Ther 308:66–72

64. Chen RJ, Ho YS, Guo HR, Wang YJ (2008) Rapid activation of Stat3 and ERK1/2 by nicotine modulates cell proliferation in human bladder cancer cells. Toxicol Sci 104:283–293

65. Stoeltzing E, Meric-Bernstam F, Ellis LM (2006) Intracellular signaling in tumor and endothelial cells: the expected and, yet again, the unexpected. Cancer Cell 10:89–91

66. Chang F, Lee JT, Navolanic PM, Steelman LS, Shelton JG, Blalock WL, Franklin RA, McCubrey JA (2003) Involvement of PI3K/Akt pathway in cell cycle progression, apoptosis, and neoplastic transformation: a target for cancer chemotherapy. Leukemia 17:590–603

67. Veschini L, Belloni D, Foglieni C, Cangi MG, Ferrarine M, Caligaris-Cappio F, Ferrero E (2007) Hypoxia-inducible transcription factor-1α determines sensitivity of endothelial cells to the proteosome inhibitor bortezomib. Blood 109:2565–2570

68. Tabruyn SP, Griffioen AW (2007) A new role for NF-κB in angiogenesis inhibition. Cell Death Differ 14:1393–1397

69. Zhang Q, Tang X, Zhang ZF, Velikina R, Shi S, Le AD (2007) Nicotine induces hypoxia-inducible factor-1α expression in human lung cancer cells via nicotinic acetylcholine receptor-mediated signaling pathways. Clin Cancer Res 13:4686–4694

70. Park SE, Shah AN, Zhang J, Gallick GE (2007) Regulation of angiogenesis and vascular permeability by Src family kinases: opportunities for therapeutic treatment of solid tumors. Expert Opin Ther Targets 11:1207–1217

71. Dasgupta P, Kinkade R, Joshi B, Decook C, Haura E, Chellappan S (2006) Nicotine inhibits apoptosis induced by chemotherapeutic drugs by up-regulating XIAP and survivin. Proc Natl Acad Sci USA 103:6332–6337

72. Jin Z, Gao F, Flagg T, Deng X (2004) Nicotine induces multi-site phosphorylation of Bad in association with suppression of apoptosis. J Biol Chem 279:23837–23844

73. Xin M, Deng X (2005) Nicotine inactivation of the proapoptotic function of Bax through phosphorylation. J Biol Chem 280:10781–10789

74. Fontanini G, Boldrini L, Vignati S, Chine S, Basolo F, Silvestri V, Lucchi M, Mussi A, Angeletti CA, Becilacqua G (1998) Bcl2 and p53 regulate vascular endothelial growth factor (VEGF)-mediated angiogenesis in non-small cell lung carcinoma. Eur J Cancer 34:718–723

75. Trombino S, Cesarioa A, Margaritora S, Granone P, Motta G, Falugi C, Russo P (2004) α7-Nicotinic acetylcholine receptors affect growth regulation of human mesothelioma cells: role of mitogen-activated protein kinase pathway. Cancer Res 64:135–145

76. Catassi A, Paleari L, Servent D, Sessa F, Dominioni L, Ognio E, Cilli M, Vacca P, Mingari M, Gaudino G, Bertino P, Paolucci M, Calcaterra A, Cesario A, Granone P, Costa R, Ciarlo M, Alama A, Russo P (2008) Targeting α7-nicotinic receptor for the treatment of pleural mesothelioma. Eur J Cancer 44:2296–2311

77. Gabellini C, Del Bufalo D, Zupi G (2006) Involvement of RB gene family in tumor angiogenesis. Oncogene 25:5326–5332

78. Murohara T, Asahara T, Silver M, Bauters C, Masuda H, Kalka C, Kearney M, Chen D, Symes JF, Fishman MC, Huang PL, Isner JM (1998) Nitric oxide synthase modulates angiogenesis in response to tissue ischemia. J Clin Invest 101:2567–2578

79. Tonnessen BH, Severson SR, Hurt RD, Miller VM (2000) Modulation of nitric-oxide synthase by nicotine. J Pharmacol Exp Ther 295:601–606

80. Sugimoto A, Masuda H, Eguchi M, Iwaguro H, Tanabe T, Asahara T (2007) Nicotine enlivenment of blood flow recovery following endothelial progenitor cell transplantation into ischemic hindlimb. Stem Cells Dev 16:649–656

81. Kimura H, Esumi H (2003) Reciprocal regulation between nitric oxide and vascular endothelial growth factor in angiogenesis. Acta Biochim Pol 50:49–59

82. Cao R, Jensen LDE, Soll R, Hauptmann G, Cao Y (2008) Hypoxia-induced retinal angiogenesis in zebrafish as a model to study retinopathy. PLoS One 3:e2748

83. Suner IJ, Espinosa-Heidmann DG, Marin-Castano ME, Hernandez EP, Pereira-Simon S, Cousins SW (2004) Nicotine increases size and severity of experimental choroidal neovascularization. Invest Ophthalmol Vis Sci 45:311–317

84. Kiuchi K, Matsuoka M, Wu JC, Lima e Silva R, Kengatharan M, Verghese M, Ueno S, Yokoi K, Khu NH, Cooke JP, Campochiaro PA (2008) Mecamylamine suppresses basal and nicotine-stimulated choroidal neovascularization. Invest Ophthalmol Vis Sci 49:1705–1711

85. Shytle RD, Penny E, Silver AA, Goldman J, Sanberg PR (2002) Mecamylamine (Inversine®): an old antihypertensive with new research directions. J Hum Hypertens 16:453–457

86. Feuerbach D, Lingenhoehl K, Olpe HR, Vassout A, Gentsch C, Chaperon F, Nozulak J, Enz A, Bilbe G, McAllister K, Hoyer D (2009) The selective nicotinic acetylcholine receptor alpha7 agonist JN403 is active in animal models of cognition, sensory gating, epilepsy and pain. Neuropharmacology 56:254–263

87. Croll SD, Goodman JH, Scharfman HE (2004) Vascular endothelial growth factor (VEGF) in seizures: a double-edged sword. Adv Exp Med Biol 548:57–68

88. Rigau V, Morin M, Rousset MC, de Bock F, Lebrun A, Coubes P, Pico MC, Baldy-Moulinier M, Bockaert J, Crespel A, Lerner-Natoli M (2007) Angiogenesis is associated with blood–brain barrier permeability in temporal lobe epilepsy. Brain 130:1942–1956

89. Gallowitsch-Puerta M, Tracey KJ (2005) Immunologic role of the cholinergic anti-inflammatory pathway and the nicotinic acetylcholine α7 receptor. Ann N Y Acad Sci USA 1062:209–219

90. Kawashima K, Yoshikawa K, Fujii YX, Moriwaki Y, Misawa H (2007) Expression and function of genes encoding cholinergic components in murine immune cells. Life Sci 80:2314–2319

91. Pavlov VA, Tracey KJ (2006) Controlling inflammation: the cholinergic anti-inflammatory pathway. Biochem Soc Trans 34:1037–1040

92. Shytle RD, Mori T, Townsend K, Vendrame M, Sun N, Zeng J, Ehrhart J, Silver AA, Sanberg PR, Tan J (2004) Cholinergic modulation of microglial activation by α7 nicotinic receptors. J Neurochem 89:337–343

93. Kindt F, Wiegand S, Niemeier V, Kupfer J, Löser C, Nilles M, Kurzen H, Kummer W, Gieler U, Haberberger RV (2008) Reduced expression of nicotinic α subunits 3, 7, 9 and 10 in lesional and nonlesional atopic dermatitis skin but enhanced expression of α subunits 3 and 5 in mast cells. Br J Dermatol 159:847–857

94. Walters EH, Reid D, Soltani A, Ward C (2008) Angiogenesis: a potentially critical part of remodeling in chronic airway diseases? Pharmacol Ther 118:128–137

95. Jackson JR, Seed MP, Kircher CH, Willoughby DA, Winkler JD (1997) The codependence of angiogenesis and chronic inflammation. FASEB J 11:457–465

96. Blanchet MR, Israel-Assayag E, Daleau P, Beaulieu MJ, Cormier Y (2006) Dimethyphenylpiperazinium, a nicotinic receptor agonist, downregulates inflammation in monocytes/macrophages through PI3K and PLC chronic activation. Am J Physiol Lung Cell Mol Physiol 291:L757–L763

97. Gwilt CR, Donnelly LE, Rogers DF (2007) The non-neuronal cholinergic system in the airways: an unappreciated regulatory role in pulmonary inflammation? Pharmacol Ther 115:208–222

98. Buitenhuis M, Verhagen LP, van Deutekom HW, Castor A, Verploegen S, Koenderman L, Jacobsen SE, Coffer PJ (2008) Protein kinase B (c-akt) regulates hematopoietic lineage choice decisions during myelopoiesis. Blood 111:112–121

99. Chng HW, Camplejohn RS, Stone MG, Hart IR, Nicholson LJ (2006) A new role for the anti-apoptotic gene A20 in angiogenesis. Exp Cell Res 312:2897–2907

100. Sainson RC, Johnston DA, Chu HC, Holderfield MT, Makatsu MN, Crampton SP, Davis J, Conn E, Hughes CC (2008) TNF primes endothelial cells for angiogenic sprouting by inducing a tip cell phenotype. Blood 111:4997–5007

101. Ferro T, Neumann P, Gertzberg N, Clements R, Johnson A (2000) Protein kinase C-α mediates endothelial barrier dysfunction inducted by TNF-α. Am J Physiol Lung Cell Mol Physiol 278:L1107–L1117

102. Doyle B, Caplice N (2007) Plaque neovascularization and antiangiogenic therapy for atherosclerosis. J Am Coll Cardiol 49:2073–2080

103. Zhen Y, Ruixing Y, Qi B, Jinzhen W (2008) Nicotine potentiates vascular endothelial growth factor expression in balloon-injured rabbit aortas. Growth Factors 26:284–292

104. Masferrer JL, Leahy KM, Koki AT, Zweifel BS, Settle SL, Woerner BM, Edwards DA, Flickinger AG, Moore RJ, Seibert K (2000) Antiangiogenic and antitumor activities of cyclo-oxygenase-2 inhibitors. Cancer Res 60:1306–1311

105. Takahashi HK, Iwagaki H, Hamano R, Yoshino T, Tanaka N, Nishibori M (2006) Effect of nicotine on IL-18-initiated immune response in human monocytes. J Leukoc Biol 80:1388–1394

106. Gahring LC, Meyer EL, Rogers SW (2003) Nicotine-induced neuroprotection against N-methyl-D-aspartic acid or β-amyloid peptide occur through independent mechanisms distinguished by pro-inflammatory cytokines. J Neurochem 87:1125–1136

107. De Simone R, Ajmone-Cat MA, Carnevale D, Minghetti L (2005) Activation of α7 nicotinic acetylcholine receptor by nicotine selectively up-regulates cyclooxygenase-2 and prostaglandin E2 in rat microglial cultures. J Neuroinflammation 2:4–14

108. Tuppo EE, Arias HR (2005) The role of inflammation in Alzheimer's disease. Int J Biochem Cell Biol 37:289–305

109. Streit WJ, Mrak RE, Griffin WS (2004) Microglia and neuroinflammation: a pathological perspective. J Neuroinflammation 1:1–14

110. Checchin D, Sennlaub F, Levavasseur E, Leduc M, Chemtob S (2006) Potential role of microglia in retinal blood vessel formation. Invest Ophthalmol Vis Sci 47:3595–3602

111. Ritter MR, Banin E, Moreno SK, Aguilar E, Dorrell MI, Friedlander M (2006) Myeloid progenitors differentiate into microglia and promote vascular repair in a model of ischemic retinopathy. J Clin Invest 116:3266–3276

112. Newell EW, Schlichter LC (2005) Integration of K$^+$ and Cl$^-$ currents regulate steady-state and dynamic membrane potentials in culture rat microglia. J Physiol 567:869–890

113. Erdogan A, Schaefer CA, Schaefer M, Luedders DW, Stockhausen F, Abdallah Y, Schaefer C, Most AK, Tillmanns H, Piper HM, Kuhlmann CR (2005) Margatoxin inhibits VEGF-induced hyperpolarization, proliferation and nitric oxide production of human endothelial cells. J Vasc Res 42:368–376

114. Singer AJ, Clark RA (1999) Cutaneous wound healing. N Engl J Med 341:738–746

115. Zia S, Ndoye A, Nguyen VT, Grando SA (1997) Nicotine enhances expression of the α3, α4, α5, and α7 nicotinic receptors modulating calcium metabolism and regulating adhesion and motility of respiratory epithelial cells. Res Commun Mol Pathol Pharmacol 97:243–262

116. Grando SA, Pittelkow MR, Schallreuter KU (2006) Adrenergic and cholinergic control in the biology of epidermis: physiological and clinical significance. J Invest Dermatol 126:1948–1965

117. Jacobi J, Jang JJ, Sundram U, Dayoub H, Fajardo LF, Cooke JP (2002) Nicotine accelerates angiogenesis and wound healing in genetically diabetic mice. Am J Pathol 161:97–104

118. Morimoto N, Takemoto S, Kawazoe T, Suzuki S (2008) Nicotine at a low concentration promotes wound healing. J Surg Res 145:199–204

119. Werner S, Krieg T, Smola H (2007) Keratinocyte-fibroblast interactions in wound healing. J Invest Dermatol 127:998–1008

120. Chernyavsky AI, Arredondo J, Vetter DE, Grando SA (2007) Central role of α9 acetylcholine receptor in coordinating keratinocyte adhesion and motility at the initiation of epithelialization. Exp Cell Res 313:3542–3555

121. Zheng LW, Ma L, Cheung LK (2008) Changes in blood perfusion and bone healing induced by nicotine during distraction osteogenesis. Bone 43:355–361

122. Feitelson JBA, Rowell PP, Roberts CS, Fleming JT (2003) Two week nicotine treatment selectively increases bone vascular constriction in response to norepinephrine. J Orthop Res 21:497–502

123. Hida K, Hida Y, Amin DN, Flint AF, Panigrahy D, Morton CC, Klagsbrun M (2004) Tumor-associated endothelial cells with cytogenetic abnormalities. Cancer Res 64:8249–8255

124. Sawada T, Kato Y, Sakayori N, Takekawa Y, Kobayashi M (1999) Expression of the multidrug-resistance P-glycoprotein (Pgp, MDR-1) by endothelial cells of the neovasculature in central nervous system tumors. Brain Tumor Pathol 16:23–27

125. Tran J, Master Z, Yu JL, Rak J, Dumont DJ, Kerbel RS (2002) A role for survivin in chemoresistance of endothelial cells mediated by VEGF. Proc Natl Acad Sci USA 99:4349–4354
126. Kem WR, Soti F, LeFrancois S, Wildeboer K, MacDougall K, Wei D, Chou KC, Arias HR (2006) The nemertine toxin anabaseine and its derivative DMXBA (GTS-21): chemical and pharmacological properties. Mar Drugs (Special Issue: Marine Toxins and Ion Channels) 4:255–273
127. Gu RX, Gu H, Xie ZY, Wang JF, Arias HR, Wei DQ, Chou KC (2009) Possible drug candidates for Alzheimer's disease deduced from studying their binding interactions with the α7 nicotinic acetylcholine receptor. Med Chem 5:250–262

Chapter 7
Catecholamine Neurotransmitters: An Angiogenic Switch in the Tumor Microenvironment

Sujit Basu and Partha Sarathi Dasgupta

Abstract Angiogenesis, or new blood vessel formation, is necessary for the growth and progression of malignant tumors. Among the endogenous regulators of angiogenesis, catecholamines have recently drawn attention owing to the discovery that they have opposing roles in regulating tumor angiogenesis. Dopamine (DA), norepinephrine (NE), and epinephrine (E) are the members of the catecholamine family. DA suppresses tumor angiogenesis and hence inhibits tumor growth, whereas NE and E increase tumor growth by promoting angiogenesis in tumor tissues. Therefore, on the whole, catecholamines function as an angiogenic switch. These neurotransmitters act upon their target cells via specific receptors, exerting pro- or anti-angiogenic effects, and thus are excellent targets for the regulation of tumor angiogenesis by dopaminergic or adrenergic receptor agonists or antagonists.

Neovascularization occurs by two distinct mechanisms: angiogenesis, in which new blood vessels sprout from the pre-existing blood vessels, and vasculogenesis, in which new blood vessels are derived from the circulating bone marrow-derived endothelial progenitor cells (EPCs) [1–5]. This process of new blood vessel formation is essential not only in normal physiological situations (e.g., the menstrual cycle, implantation, embryogenesis, and wound-healing), but also in the growth and metastasis of malignant tumors [1–5].

In the normal physiological milieu, angiogenesis is tightly regulated by intricately balanced endogenous pro- and anti-angiogenic molecules [6–8]. However, in malignancy, this fine tuning of the balance between pro- and anti-angiogenic

S. Basu (✉)
Department of Pathology, Arthur G. James Comprehensive Cancer Center,
Ohio State University, Columbus, OH, USA
e-mail: sujit.basu@osumc.edu

P.S. Dasgupta
Signal Transduction and Biogenic Amines Department,
Chittaranjan National Cancer Institute, Kolkata, India

S.A. Mousa and P.J. Davis (eds.), *Angiogenesis Modulations in Health and Disease:*
Practical Applications of Pro- and Anti-angiogenesis Targets,
DOI 10.1007/978-94-007-6467-5_7, © Springer Science+Business Media Dordrecht 2013

molecules is disabled by either the overexpression of pro-angiogenic molecules or the down-regulation of anti-angiogenic molecules, resulting in the activation of the angiogenic switch [6–8]. The overexpression of pro-angiogenic growth factors, such as vascular permeability factor/vascular endothelial growth factor (VPF/ VEGF), fibroblast growth factor (FGF), interleukin-8 (IL-8), placenta growth factor (PlGF), transforming growth factor β (TGF-β), and platelet derived growth factor (PDGF) tilt the tumor microenvironment in favor of angiogenesis and allow the transition of an avascular dormant tumor into a growing vascular tumor mass [1–3]. Because targeting growth factor-induced angiogenesis has shown clinical promise, designing therapies to target tumor neovessels is of interest owing to the decreased toxicity of the approach, the minimal drug resistance, and the ability to increase the efficacies of anti-cancer drugs and radiation therapy [1–10].

Catecholamines are a group of neurotransmitters that includes dopamine (DA), norepinephrine (NE), and epinephrine (E) [11]. In addition to their conventional roles in the brain, these molecules also have important functions in the periphery [11]. Recent discoveries of a regulatory role for different catecholamines in tumor angiogenesis are of current interest from a clinical viewpoint for the development of anti-angiogenic drugs to treat cancer patients [12–14]. These newly identified roles for catecholamines also enable us to understand the biology of catecholamines in peripheral systems [13]. The available information regarding the roles of catechol-amines in the regulation of tumor angiogenesis is discussed in this chapter.

NE and E are Endogenous Promoters of Tumor Angiogenesis

NE and E act on their target cells through α (α_1 and α_2) and β (β_1 and β_2) adrenoceptors [11]. Evidence indicates that exposure to chronic stress promotes tumor growth [12, 15] through the stress mediators NE and E [16, 17], and the up-regulation of tumor angiogenesis is suggested to be the underlying mechanism [13, 18]. In a model of orthotopically xenografted human ovarian tumors in nude mice, a tumor growth-promoting effect was observed in animals following exposure to chronic stress or treatment with the β-adrenergic agonist isoproterenol, and this effect was abrogated by the β-adrenergic antagonist propranolol [19]. Interestingly, this increase in tumor growth was associated with the up-regulation of VEGF in tumor tissues, which led to the induction of tumor angiogenesis [19]. However, inhibition of the VEGF pathway suppressed the tumor growth stimulatory effect of the β-adrenergic agonist [19]. *In vitro* studies also demonstrated the NE-mediated secretion of VEGF by ovarian carcinoma cells [20]. In addition, there are reports that indicate that NE, by acting on the adreno-ceptors present in the tumor-associated macrophages (TAM), induces angiogenesis by stimulating the production of matrix metalloproteinase 9 (MMP-9) [21].

In addition to the NE-mediated increase in the expression of the pro-angiogenic cytokine VEGF, studies have also indicated that in different tumor cells bearing the β-adrenoceptor (such as melanoma, ovarian, and nasopharyngeal cancer cells), NE induces a significant increase in the synthesis and release of other pro-angiogenic factors, including IL-6, IL-8, MMP-2, and MMP-9 [21–24]. Interestingly, nicotine has been shown to increase xenografted human colon tumor growth in nude mice. This increased tumor growth is associated with elevated plasma E levels and tumor angiogenesis. However, blocking the β-adrenoceptors with specific antagonists significantly abrogated the nicotine-induced tumor growth through the down-regulation of tumor angiogenesis [25]. In addition, NE has also been shown to stimulate VEGF mRNA synthesis in endothelial cells through the cAMP-PKA pathway and to promote proliferation of these cells by activating ERK [26].

Molecular Mechanisms of NE- and E-induced Tumor Angiogenesis

By acting through β_2-adrenoceptors, NE has been shown to promote angiogenesis in the orthotopically grown ovarian cancers HEY-8 and SKOV3ipI [19]. The underlying molecular mechanisms of this phenomenon were determined to be increased VEGF synthesis and the overexpression of matrix metalloproteinases, such as MMP-2 and MMP-9 [19]. Further investigations demonstrated that this VEGF-induced overexpression of matrix metalloproteinases is mediated through the cAMP-PKA signaling pathway following stimulation of β_2-adrenoceptors by NE, indicating a novel signaling pathway, such as NE-β_2-adrenoceptors-cAMP-PKA-VEGF [19]. NE treatment has also shown similar results in the human nasopharyngeal cell line HONE 1 [23]. In several human multiple myeloma cell lines (NCI-H-929, MM-M1, and FLAM-76) NE treatment has also shown similar results by acting through β_1 and β_2 adrenoceptors present in these cells [27]. However, a recent study has demonstrated that *in vitro* treatment of human prostate (PC3), breast (MDA-MB-231), and liver (HCC SK-Hep1) cancer cells with NE or isoproterenol stimulated the expression of HIF-1α and synthesis of VEGF in a dose-dependent manner [28]. This increased VEGF synthesis was decreased when the tumor cells were transfected with HIF-1α siRNA [28]. This observation was further strengthened when HIF-1α was up- or down-regulated in these tumor cells following pretreatment with the adenylate cyclase activator forskolin or the protein kinase A (PKA) inhibitor H-89, respectively [28]. Finally, pretreatment of tumor cells with a β-adrenergic blocker propranolol completely abolished the expression of VEGF and HIF-1α protein amount in these cells [28]. Therefore, in brief, NE induces VEGF expression in cancer cells through NE-β-adrenoceptor-PKA-HIF-1α-VEGF signaling pathway (Fig. 7.1).

This catecholamine neurotransmitter also stimulates the synthesis and release of another pro-angiogenic factor, IL-6, in the human ovarian tumor cell lines SKOV3ip1, HEY-A8 and EG *in vitro* [24]. By acting through β-adrenoceptors in these tumor cells, NE significantly increased both IL-6 mRNA synthesis and promoter activity [24]. Additional results have demonstrated an abrogation of this NE-mediated effect on IL-6 synthesis following treatment with β-adrenoceptor antagonists, confirming the NE-mediated regulation of IL-6 gene transcription through the activation of β-adrenoceptors [24]. NE-mediated β-adrenoceptor activation was also shown to increase Src kinase phosphorylation, which subsequently increased IL-6 mRNA synthesis through the up-regulation of the IL-6 promoter activity [24]. This suggestion of a NE-β-adrenoceptor-Src kinase-IL-6 pathway for

Fig. 7.1 NE stimulates VEGF synthesis in the tumor cells. NE by activating β-adrenergic receptors activates cAMP-PKA axis and stimulates VEGF synthesis by up-regulating the transcription factor Hypoxia Inducible Factor-1α (HIF-1α) in tumor cells

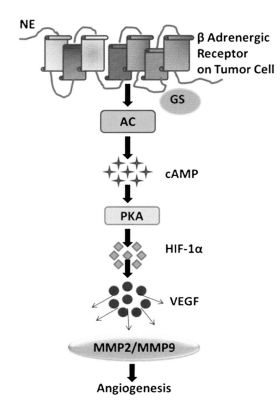

increased tumor angiogenesis was further strengthened by immunohistochemical analysis of human ovarian cancer tissues, in which a significant correlation between the overexpression of Src kinase and the degree of tumor neovascularization was observed [24]. Src activation was also instrumental in increasing the synthesis of other pro-angiogenic molecules, such as VEGF and IL-8 [24]. However, another alternate signaling pathway was recently identified in which NE and E stimulated MMPs in human ovarian tumor cell lines independent of the β1, β2-adrenoceptors-PKA pathway. This pathway involves STAT-3, a transcription factor known to initiate several signaling pathways in cancer cells [29].

DA as an Endogenous Inhibitor of Tumor Angiogenesis

In addition to acting as a precursor molecule in the biosynthetic pathway of NE and E, DA also acts as an important neurotransmitter in both the brain and the peripheral organs [30]. In the brain, DA regulates several major functions, including cognition, motor activities, and the reward effect in the form of pleasure [30, 31]. In peripheral systems, DA regulates cardiac and renal functions. In addition, recent evidence has indicated that DA influences other diverse functions, such as blood pressure, insulin synthesis in beta cells of the pancreas, and the functions of immune effector cells [32–34]. Recently, another role of this neurotransmitter in the peripheral system has been demonstrated: DA functions as an endogenous inhibitor of angiogenesis by acting through its D_2 class of receptors present in the endothelial cells and EPCs [35–41].

A significant increase in B-16 melanoma growth has been found in D_2 DA receptor ($-/-$) mice [37]. Another study has revealed significantly decreased growth of mammary carcinoma in hyperdopaminergic Wistar rats (APO-SUS), and increased growth of the same tumors was observed in hypodopaminergic rats (APO-UNSUS) [36]. This increased or decreased tumor growth in hypo- or hyperdopamineregic rats is closely associated with increased or decreased angiogenesis in tumors [36]. Furthermore, in human gastric cancer patients, a significant reduction of DA in malignant stomach tissues compared to the surrounding normal tissues has been observed, and exogenous administration of DA or a D_2 DA receptor agonist significantly inhibited stomach tumor growth [38]. The mechanism of this phenomenon has been attributed to the inhibition of angiogenesis in the tumor tissues [38].

Molecular Mechanisms of DA-induced Inhibition of Tumor Angiogenesis

VEGF is the predominant cytokine that regulates angiogenesis by mediating proliferation, migration, and tube formation in endothelial cells from pre-existing vessels [3, 10]. VEGF also plays a pivotal role in the migration and subsequent mobilization of EPCs from the bone marrow into the neovessels of tumors by acting through VEGFR-2 receptors present on these cells [41]. *In vivo* studies have demonstrated that DA treatment significantly inhibits tumor angiogenesis [35–41]. Tumor endothelial cells isolated from human breast (MCF-7) and colon (HT29) tumor-bearing mice displayed suppression of VEGFR-2 phosphorylation with subsequent inhibition of its downstream signaling cascades (e.g., MAPK and focal adhesion kinase (FAK)), which regulate the proliferation and migration of endothelial cells; this regulation is essential for tumor neovessel formation (Fig. 7.2) [40]. Recent studies also reveal important contributions of bone marrow-derived EPCs in tumor angiogenesis [4, 5, 42]. Indeed, additional reports indicate that in the bone marrow niche, DA is synthesized in stromal cells [43] and is depleted in tumor-bearing mice [41], thus indicating a role of DA in the regulation of the mobilization of these precursor cells from bone marrow into circulation [41]. Importantly, the administration of exogenous DA, which inhibited tumor angiogenesis, also inhibited the mobilization of these cells from bone marrow [41]. This inhibitory effect of DA is abrogated when the animals are treated with a D_2 DA receptor antagonist [41]. The inhibitory effect on the mobilization of EPCs from the bone marrow and hence on tumor angiogenesis is due to the D_2 DA receptor-mediated down-regulation of MMP-9 synthesis by these bone marrow progenitor cells through the inhibition of the ERK1/ERK2 pathway (Fig. 7.3) [41]. These observations were further supported in D_2 DA receptor ($-/-$) mice: increased numbers of circulating EPCs

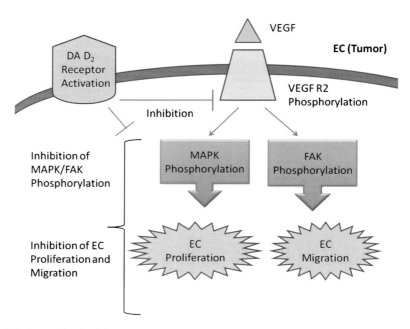

Fig. 7.2 Dopamine inhibits tumor endothelial cell proliferation and migration. Dopamine by activating its D$_2$ receptors inhibits VEGFR-2 phosphorylation and downstream signaling molecules like mitogen-activated protein kinase (MAPK) and Focal adhesion kinase (FAK)

were evident in tumor-bearing mice compared to wild type controls, and the D$_2$ DA receptor antagonist treatment failed to elicit any effect in the animals [41].

These studies have clearly demonstrated that DA acts as an endogenous inhibitor of angiogenesis, and hence of tumor growth, by targeting the VEGF-induced proliferation and migration of endothelial cells and EPCs through several newly uncovered mechanisms [35–41].

Discussion

Together, current findings show that catecholamine neurotransmitters act as regulators of tumor angiogenesis and, hence, regulate the growth of malignant tumors [13]. The available evidence suggests that DA acts through its D$_2$ class of receptors to inhibit tumor angiogenesis by targeting the proliferation and migration of tumor endothelial cells as well as the mobilization of EPCs [35–41]. In contrast, NE and E act through β-adrenoceptors to stimulate angiogenesis by stimulating the synthesis of pro-angiogenic cytokines (e.g., VEGF, IL-8, and IL-6) and MMPs (e.g., MMP-2 and MMP-9) in tumor cells through different signaling pathways [13]. Briefly, in the D$_2$ DA receptor-mediated down-regulation of angiogenesis, the target cells are endothelial cells and EPCs, whereas in the NE-mediated up-regulation of tumor angiogenesis, the target cells are principally tumor cells [13]. Therefore, based on these opposing

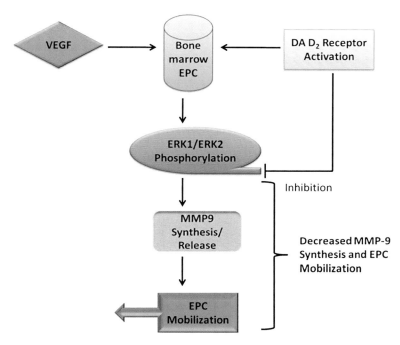

Fig. 7.3 Dopamine inhibits mobilization of bone marrow-derived endothelial progenitor cells for tumor neovessel formation. Activation of D_2 dopamine receptors in endothelial progenitor cells (EPCs), dopamine inhibits migration of these cells from bone marrow to the circulation and subsequently to neovessels of the tumors by inhibiting synthesis of matrix metalloproteinase-9 (MMP-9) in these progenitor cells

Fig. 7.4 Model of an angiogenic switch in tumor: Diagram of catecholamine-mediated operation of an angiogenic switch in tumor microenvironment. Dopamine inhibits angiogenesis, whereas norepinephrine and epinephrine up-regulate angiogenesis in tumor tissues

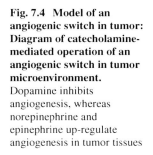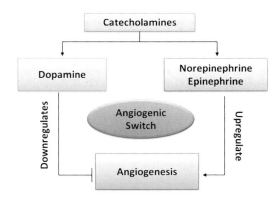

effects of stimulation and inhibition of tumor angiogenesis by the catecholamine neurotransmitters (DA, NE, and E), it is suggested that catecholamines function as an angiogenic switch in the tumor microenvironment (Fig. 7.4). The expression profile of D_2 DA receptors or β-adrenoceptors in any organ may be altered in response to the onset of malignancy in that organ. These alterations may tilt the microenvironment of the tumor in favor of angiogenesis, thereby transforming an avascular, dormant tumor mass into a vascular, rapidly growing tumor. Epidemiological evidence has also indicated that the use of NE antagonists reduces the risk of cancer incidence [44]. Therefore, it will be prudent to undertake further detailed investigations to dissect the specific roles of catecholamines in relation to their function as an angiogenic switch in tumor growth. These studies will enable clinicians to develop DA or NE/E receptor agonists or antagonists as anti-angiogenic drugs.

References

1. Folkman J, Shing Y (1992) Angiogenesis. J Biol Chem 267:10931–10934
2. Carmeliet P, Jain RK (2000) Angiogenesis in cancer and other diseases. Nature 407:249–257
3. Dvorak HF (2005) Angiogenesis: update 2005. J Thromb Haemost 3:1835–1842
4. Asahara T, Murohara T, Sullivan A et al (1997) Isolation of putative progenitor endothelial cells for angiogenesis. Science 275:964–967
5. Kopp HG, Ramos CA, Rafii S (2006) Contribution of endothelial progenitors and proangiogenic hematopoietic cells to vascularization of tumor and ischemic tissue. Curr Opin Hematol 13:175–181
6. Moserle L, Amadori A, Indraccolo S (2009) The angiogenic switch: implications in the regulation of tumor dormancy. Curr Mol Med 9:935–941
7. Naumov GN, Akslen LA, Folkman J (2006) Role of angiogenesis in human tumor dormancy: animal models of the angiogenic switch. Cell Cycle 5:1779–1787
8. Baeriswyl V, Christofori G (2009) The angiogenic switch in carcinogenesis. Semin Cancer Biol 19:329–337
9. Cai J, Han S, Qing R et al (2011) In persuit of new anti-angiogenic therapies for cancer treatment. Front Biosci 16:803–814
10. Ferrara N (2009) Vascular endothelial growth factor. Arterioscler Thromb Vasc Biol 29:789–791
11. Laverty R (1978) Catecholamines: role in health and disease. Drugs 16:418–440
12. Antoni MH, Lutgendorf SK, Cole SW et al (2006) The influence of bio-behavioral factors on tumor biology: pathways and mechanisms. Nat Rev Cancer 6:240–248
13. Chakroborty D, Sarkar C, Basu B et al (2009) Catecholamines regulate tumor angiogenesis. Cancer Res 69:3727–3730
14. Tilan J, Kitlinska J (2010) Sympathetic neurotransmitters and tumor angiogenesis-link between stress and cancer progression. J Oncol 2010:539706. doi:10.1155/2010/539706
15. Hasegawa H, Saiki I (2002) Psychosocial stress augments tumor development through beta-adrenergic activation in mice. Jpn J Cancer Res 93:729–735
16. Thaker PH, Sood AK (2008) Neuroendocrine influence on cancer biology. Semin Cancer Biol 18:164–170
17. Thaker PH, Lutgendorf SK, Sood AK (2007) The neuroendocrine impact of chronic stress on cancer. Cell Cycle 6:430–433
18. Armaiz-Pena GN, Lutgendorf SK, Cole SW, Sood AK (2009) Neuroendocrine modulation of cancer progression. Brain Behav Immun 23:10–15
19. Thaker PH, Han LY, Kamat AA et al (2006) Chronic stress promotes tumor growth and angiogenesis in a mouse model of ovarian carcinoma. Nat Med 12:939–944
20. Lutgendorf SK, Cole S, Costanzo E et al (2003) Stress-related mediators stimulate vascular endothelial growth factor secretion by two ovarian cancer cell lines. Clin Cancer Res 9:4514–4521

21. Lutgendorf SK, Lamkin DM, Jennings NB (2008) Biobehavioral influences on matrix metalloproteinase expression in ovarian carcinoma. Clin Cancer Res 14:6839–6846
22. Yang EV, Kim SJ, Donovan EL et al (2009) Norepinephrine upregulates VEGF, IL-8, and IL-6 expression in human melanoma tumor cell lines: implications for stress-related enhancement of tumor progression. Brain Behav Immun 23:267–275
23. Yang EV, Sood AK, Chen M et al (2006) Norepinephrine up-regulates the expression of vascular endothelial growth factor, matrix metalloproteinase (MMP)-2, and MMP-9 in nasopharyngeal carcinoma tumor cells. Cancer Res 66:10357–10364
24. Nilsson MB, Armaiz-Pena GN, Takahashi R et al (2007) Stress hormones regulate interleukin-6 expression by human ovarian carcinoma cells through a Src-dependent mechanism. J Biol Chem 282:29919–29926
25. Wong HP, Yu L, Lam EK et al (2007) Nicotine promotes colon tumor growth and angiogenesis through beta-adrenergic activation. Toxicol Sci 97:279–287
26. Seya Y, Fukuda T, Isobe K et al (2006) Effect of norepinephrine on RhoA, MAP kinase, proliferation and VEGF expression in human umbilical vein endothelial cells. Eur J Pharmacol 553:54–60
27. Yang EV, Donovan EL, Benson DM, Glaser R (2008) VEGF is differentially regulated in multiple myeloma-derived cell lines by norepinephrine. Brain Behav Immun 22:318–323
28. Park SY, Kang JH, Jeong KJ et al (2011) Norepinephrine induces VEGF expression and angiogenesis by a hypoxia-inducible factor-1α protein-dependent mechanism. Int J Cancer 128:2306–2316
29. Landen CN Jr, Lin YG, Armaiz Pena GN et al (2007) Neuroendocrine modulation of signal transducer and activator of transcription-3 in ovarian cancer. Cancer Res 67:10389–10396
30. Beaulieu JM, Gainetdinov RR (2011) The physiology, signaling, and pharmacology of dopamine receptors. Pharmacol Rev 63:182–217
31. Missale C, Nash SR, Robinson SW, Jaber M, Caron MG (1998) Dopamine receptors: from structure to function. Physiol Rev 78:189–225
32. Rubi B, Maechler P (2010) Minireview: new roles for peripheral dopamine on metabolic control and tumor growth: let's seek the balance. Endocrinology 151:5570–5581
33. Basu S, Dasgupta PS (2000) Dopamine, a neurotransmitter, influences the immune system. J Neuroimmunol 102:113–124
34. Sarkar C, Basu B, Chakroborty D et al (2010) The immunoregulatory role of dopamine: an update. Brain Behav Immun 24:525–528
35. Basu S, Nagy JA, Pal S et al (2001) The neurotransmitter dopamine inhibits angiogenesis induced by vascular permeability factor/vascular endothelial growth factor. Nat Med 7:569–574
36. Teunis MA, Kavelaars A, Voest E et al (2002) Reduced tumor growth, experimental metastasis formation, and angiogenesis in rats with a hyperreactive dopaminergic system. FASEB J 16:1465–1467
37. Basu S, Sarkar C, Chakroborty D et al (2004) Ablation of peripheral dopaminergic nerves stimulates malignant tumor growth by inducing vascular permeability factor/vascular endothelial growth factor-mediated angiogenesis. Cancer Res 64:5551–5555
38. Chakroborty D, Sarkar C, Mitra RB et al (2004) Depleted dopamine in gastric cancer tissues: dopamine treatment retards growth of gastric cancer by inhibiting angiogenesis. Clin Cancer Res 10:4349–4356
39. Sarkar C, Chakroborty D, Mitra RB et al (2004) Dopamine in vivo inhibits VEGF-induced phosphorylation of VEGFR-2, MAPK, and focal adhesion kinase in endothelial cells. Am J Physiol Heart Circ Physiol 287:H1554–H1560
40. Sarkar C, Chakraborty D, Chowdhury UR et al (2008) Dopamine increases the efficacy of anticancer drugs in breast and colon cancer preclinical models. Clin Cancer Res 14:2502–2510
41. Chakroborty D, Chowdhury UR, Sarkar C et al (2008) Dopamine regulates endothelial progenitor cell mobilization from mouse bone marrow in tumor vascularization. J Clin Invest 118:1380–1389

42. Gao D, Nolan D, McDonnell K et al (2009) Bone marrow-derived endothelial progenitor cells contribute to the angiogenic switch in tumor growth and metastatic progression. Biochim Biophys Acta 1796:33–40
43. Marino F, Cosentino M, Bombelli R et al (1997) Measurement of catecholamines in mouse bone marrow by means of HPLC with electrochemical detection. Haematologica 82:392–394
44. Friedman GD, Udaltsova N, Habel LA (2011) Norepinephrine antagonists and cancer risk. Int J Cancer 128:737–738

Chapter 8
Impact of Nanotechnology on Therapeutic Angiogenesis

Dhruba J. Bharali and Shaker A. Mousa

Abstract Controlled and stipulated induction of angiogenesis, commonly known as therapeutic angiogenesis, can be used as a treatment strategy for many ischemic diseases and tissue regenerations. The current and conventional method of therapeutic angiogensis, which takes advantage of this natural phenomenon by inducing neovascularization in a specific body organ or site, has many disadvantages including side effects and higher dose requirements. The emergence of nanotechnology, particularly its use as a multifunctional and surface-tunable nanocarrier with site specificity and the ability to incorporate and protect active materials from *in vivo* environmental assault, has the potential for a sea change in the therapeutic angiogenesis area of research. This chapter gives a brief description of some of the nanotechnology-based therapeutic angiogenesis strategies used in recent years. Though most of the strategies, which involve delivering growth factors and therapeutic genetic materials, are in a preclinical state, there are great potential clinical implications for the near future.

Introduction

Angiogenesis is an indispensable physiological process of the formation and propagation of new blood vessels from pre-existing vasculatures, and it is the building block of adult fundamental development [1]. Lack of angiogenesis or insufficient angiogenesis is one of the major causes behind many ischemic diseases [2]. Controlled and stipulated induction of angiogenesis, popularly known as therapeutic angiogenesis, takes advantage of this natural phenomenon by inducing neovascularization in a

D.J. Bharali • S.A. Mousa (✉)
The Pharmaceutical Research Institute at Albany College of Pharmacy
and Health Sciences, Rensselaer, NY, USA
e-mail: dhruba.bharali@acphs.edu; shaker.mousa@acphs.edu

S.A. Mousa and P.J. Davis (eds.), *Angiogenesis Modulations in Health and Disease:*
Practical Applications of Pro- and Anti-angiogenesis Targets,
DOI 10.1007/978-94-007-6467-5_8, © Springer Science+Business Media Dordrecht 2013

specific body organ or site and can be used as a treatment strategy for many ischemic diseases and tissue regenerations [2–5]. Despite the tremendous effort in recent years via various therapies to treat these diseases, ischemic diseases remain one of the major causes of death worldwide. Currently, therapeutic angiogenesis mainly uses various growth factors like vascular endothelial growth factor (VEGF) for revascularization [6, 7]. However, in clinical trials it was reported that a low dose of VEGF is highly ineffective, and the therapeutic benefit of VEGF can only be obtained at high doses [8].

There are other limited approaches to promote angiogenesis that include genetically enhanced stem cells and siRNA/DNA [3, 9–11]. However, the development of delivery vehicles that can effectively deliver either pro-angionesis factors like VEGF or genetic material is still a major challenge. Nanotechnology, which uses a submicron particulate carrier system, might be the answer to the problems that scientists are facing in therapeutic angiogenesis. Rapid emergence of nanotechnology in the biomedical field in recent years has significantly impacted treatment and diagnosis of various diseases. Several nanotechnology-based drugs came on the market within a short span of time [12, 13]. The rapid emergence of nanoparticulate carrier systems in the drug delivery field, including therapeutic angiogenesis, is attributed to factors such as multifunctional and surface-tunable properties of the nanoparticles, site-specific delivery, incorporation and protection of active materials from *in vivo* environmental assault, controlled release, and increased bioavailability of incorporated active material [14–17].

This chapter gives a brief description of some of the nanotechnology-based therapeutic angiogenesis strategies used in recent years. Though most strategies involving delivery of growth factors and therapeutic genetic materials are in a preclinical state, there is great potential clinical implication for this nanotechnology in the near future.

Nanoparticles-mediated Delivery of VEGF

VEGF is one of the most commonly used pro-angiogenesis agents and has been widely used in both clinical and preclinical studies [18]. However, the poor clinical outcome of this agent (and need of high dose for required effects) have compelled the research community either to look for new alternatives or to reformulate this agent to enhance the efficacy. It has been postulated that the poor clinical outcome of VEGF might be due to its short half-life, lack of specificity, and the higher dose requirement. Nanotechnology has proven to be the ideal technology that has the potential not only to increase the half-life of the drug, but also to protect the drug from enzymatic degradation as well as provide a slow, sustained release.

In a study by Golub et al., VEGF was encapsulated in PLGA (NPs) by double emulsion solvent evaporation methods [8]. These injectable, nanoparticulated formulations with an average particle size of about 400 nm were able to release VEGF over 4 days in a sustained way. The nanoformulation was tested in a mouse fermoral artery ischemia model (129/Sv mice) and compared with free VEGF. It was observed that the mice treated with VEGF-NPs showed a significant increase in total vessel volume and vessel connectivity when compared with free VEGF and saline [8]. The vascular formation was mainly measured by microCT angiography, and this method was further validated with traditional methods like immunohistochemistry. In terms of vessel density or spacing and anisotropy (spatial orientation of vessels), there were no statistically significant differences found between the VEGF-NPs and VEGF. Though these nanoformulations showed potential in delivering VEGF effectively, the nanofabrication methods, low loading capacity, and use of a toxic organic solvent are still major hurdles that need to be addressed in the near future in order to move forward with this formulation [8].

In an interesting study by Kim et al. [19], a unique approach was used to target and deliver VEGF in a murine model of hind limb ischemia (a common model of peripheral artery diseases (PAD)), by taking advantage of effects of a phenomenon known as EPR (enhanced permeability and retention). EPR takes advantage of leaky vascularizations due to excessive secretion of angiogenic growth factor. Pegylated silica nanoparticles (SiNPs) were used to deliver VEGF to enhance the angiogenic process in the murine model. It was hypothesized that pegylated NPs carrying VEGF can take advantage of the EPR effect to enhance its accumulation in the ischemic tissue site, where sufficient presence of growth factor is a prerequisite to enhance angiogenesis to support ischemic tissue recovery. The concept of targeted delivery was proved by injecting dye-labeled (cy5.5) pegylated SiNPs (PEG-Cy-SiNPs) and nonpegylated NPs in the hind limb ischemia model. Higher levels of PEG-Cy-SiNPs were found in ischemic tissues compared to normal tissue. It also showed that NPs labeled with a dye for fluorescent imaging can be used as an imaging probe to track their site-specific accumulation in ischemic tissues. On the other hand, nonpegylated NPs were rapidly cleared by the reticulo-endothelial system (RES), including liver and spleen. In the same study they demonstrated enhanced accumulations of gold nanoparticles conjugated to VEGF in ischemic muscle tissues with a 2-fold increase of nano VEGF compared to nonconjugated VEGF [19]. This study demonstrates the potential of nanotechnology-mediated targeted delivery and imaging, which can be used for therapeutic angiogenesis.

Nanocarriers and Myocardial Ischemia

Myocardial ischemia (MI) is one of the most common causes of many heart diseases that can ultimately lead to more severe consequences like a heart attack [20]. Conventional MI therapy includes highly invasive treatments like heart surgery, transplantation, and the use of various implantable devices. Numerous medications are used to treat MI, however, dose limitation, poor half-life period, and accessibility of the drug to the ischemic myocardium are a few of the major obstacles in the treatment of MI. Therefore it is imperative to look for nonconventional systems like a nanocarrier system that has the potential to effectively treat such diseases. In a study by Binsalamah et al. [20], the potential of chitosan-alginate NPs to treat MI in a rat model was demonstrated. They synthesized NPs incorporating placental growth factor (PlGF), a reported angiogenesis stimulator, and assessed the efficacy of the NPs in an acute MI model in rats after intramyocardial injection. A significant increase in left ventricular function and vascular density were reported 8 weeks after coronary ligation by nanoparticulate PlGF, and thus this was a clear indication of improved efficacy of nanoformulated PlGF. Though intramyocardial injection is also considered to be a highly invasive technique compared to oral and systemic delivery routes, the authors suggested that modification or reformulation of the nanoformulation for a less invasive systemic delivery will bring about a potential treatment revolution in the near future [20].

Engineered Nanocarriers

Engineered nanocarrier systems are also an attractive mode of delivering various pro-angiogenic proteins in a sustained way to the pathological site by maintaining the integrity of the protein and the pro-angiogenic activity of the incorporated materials. In a study by Roy et al. [15], a nanoformulation made up of PLGA NPs encapsulating a nonglycosylated active fragment of hepatocyte growth factor/scatter factor, 1K1 (reported to be a potent angiogenic agent), was synthesized. It was found that these NPs were able to release 1K1 in a sustained way, which can temporally enable downstream signaling through mitogen-activated kinase pathways (MAPK). This *in vitro* and *in vivo* demonstration of NPs mediated technology altering MAPK signaling pathways via controlled release of growth factor for an enhanced angiogenic result has opened a new door to inducing neovascularization under impaired ischemic conditions.

Another cutting-edge approach that has the potential for a paradigm shift in therapeutic angiogenesis is the combination of genetically engineered stem cells and nanotechnology to promote angiogenesis. Though in recent years there have been several reports using stem cells to promote angiogenesis, the clinical applications remain limited due to insufficient expression of angiogenic factors, logistics, and economics. The use of NPs made of biodegradable polymers and encoded genetic material to engineer stem cells to efficiently express angiogenic factors is another potential alternative and has tremendous potential in treating many of the ischemic diseases (including MI and cerebral ischemia). In a study by Yang et al. [21], poly β-amino ester-based NPs incorporating *hVEGF* gene were synthesized and delivered effectively to human mesenchymal stem cells (hMSC) and human embryonic stem cell-derived cells (hESdCs). When tested in this genetically engineered stem cell mouse ischemic hind limb model, 2- to 4-fold higher vessel densities were reported compared to the lipofectamine transfected cells. Enhanced angiogenesis and limp salvage was observed after the intramuscular injection of hMSC transfected *VEGF* after 4 weeks of injections. Thus, the combination of nanotechnology and stem cell engineering might be able to be used to effectively treat ischemic diseases.

Future Perspective

Nanotechnology has made a sea change in every aspect of human life, from stain-free fabrics to cosmetics, rocket science, and medicine. Applications of nanotechnology have contributed to the progress in cancer research and diagnosis and are compelling the medical research community to think 'out of the box'. There have been some efforts to use nanotechnology as a delivery vehicle for the development of highly effective modalities to treat and stimulate revascularization of ischemic tissue in a preclinical setting. Though the preliminary results are promising, more systematic research is needed from the medical community to make this cutting-edge technology become a clinical reality.

References

1. Freedman SB, Isner JM (2001) Therapeutic angiogenesis for ischemic cardiovascular disease. J Mol Cell Cardiol 33:379–393
2. Griffioen AW, Molema G (2000) Angiogenesis: potentials for pharmacologic intervention in the treatment of cancer, cardiovascular diseases, and chronic inflammation. Pharmacol Rev 52:237–268
3. Losordo DW, Dimmeler S (2004) Therapeutic angiogenesis and vasculogenesis for ischemic disease: part II: cell-based therapies. Circulation 109:2692–2697
4. Ouma GO, Jonas RA, Usman MH, Mohler ER 3rd (2012) Targets and delivery methods for therapeutic angiogenesis in peripheral artery disease. Vasc Med 17:174–192
5. Deveza L, Choi J, Yang F (2012) Therapeutic angiogenesis for treating cardiovascular diseases. Theranostics 2:801–814
6. Takeshita S, Pu LQ, Stein LA, Sniderman AD, Bunting S, Ferrara N, Isner JM, Symes JF (1994) Intramuscular administration of vascular endothelial growth factor induces dose-dependent collateral artery augmentation in a rabbit model of chronic limb ischemia. Circulation 90:II228–II234
7. Takeshita S, Zheng LP, Brogi E, Kearney M, Pu LQ, Bunting S, Ferrara N, Symes JF, Isner JM (1994) Therapeutic angiogenesis. A single intraarterial bolus of vascular endothelial growth factor augments revascularization in a rabbit ischemic hind limb model. J Clin Invest 93:662–670
8. Golub JS, Kim YT, Duvall CL, Bellamkonda RV, Gupta D, Lin AS, Weiss D, Robert Taylor W, Guldberg RE (2010) Sustained VEGF delivery via PLGA nanoparticles promotes vascular growth. Am J Physiol Heart Circ Physiol 298:H1959–H1965
9. Isner JM, Walsh K, Symes J, Pieczek A, Takeshita S, Lowry J, Rosenfield K, Weir L, Brogi E, Jurayj D (1996) Arterial gene transfer for therapeutic angiogenesis in patients with peripheral artery disease. Hum Gene Ther 7:959–988
10. Liew A, O'Brien T (2012) Therapeutic potential for mesenchymal stem cell transplantation in critical limb ischemia. Stem Cell Res Ther 3:28
11. Takeshita S, Weir L, Chen D, Zheng LP, Riessen R, Bauters C, Symes JF, Ferrara N, Isner JM (1996) Therapeutic angiogenesis following arterial gene transfer of vascular endothelial growth factor in a rabbit model of hindlimb ischemia. Biochem Biophys Res Commun 227:628–635
12. Bharali DJ, Mousa SA (2010) Emerging nanomedicines for early cancer detection and improved treatment: current perspective and future promise. Pharmacol Ther 128:324–335
13. Bharali DJ, Khalil M, Gurbuz M, Simone TM, Mousa SA (2009) Nanoparticles and cancer therapy: a concise review with emphasis on dendrimers. Int J Nanomedicine 4:1–7

14. Moldovan NI, Ferrari M (2002) Prospects for microtechnology and nanotechnology in bioengineering of replacement microvessels. Arch Pathol Lab Med 126:320–324
15. Sinha Roy R, Soni S, Harfouche R, Vasudevan PR, Holmes O, de Jonge H, Rowe A, Paraskar A, Hentschel DM, Chirgadze D, Blundell TL, Gherardi E, Mashelkar RA, Sengupta S (2010) Coupling growth-factor engineering with nanotechnology for therapeutic angiogenesis. Proc Natl Acad Sci U S A 107:13608–13613
16. Kubo M, Egashira K, Inoue T, Koga J, Oda S, Chen L, Nakano K, Matoba T, Kawashima Y, Hara K, Tsujimoto H, Sueishi K, Tominaga R, Sunagawa K (2009) Therapeutic neovascularization by nanotechnology-mediated cell-selective delivery of pitavastatin into the vascular endothelium. Arterioscler Thromb Vasc Biol 29:796–801
17. Sajja HK, East MP, Mao H, Wang YA, Nie S, Yang L (2009) Development of multifunctional nanoparticles for targeted drug delivery and noninvasive imaging of therapeutic effect. Curr Drug Discov Technol 6:43–51
18. Ferrara N, Alitalo K (1999) Clinical applications of angiogenic growth factors and their inhibitors. Nat Med 5:1359–1364
19. Kim J, Cao L, Shvartsman D, Silva EA, Mooney DJ (2011) Targeted delivery of nanoparticles to ischemic muscle for imaging and therapeutic angiogenesis. Nano Lett 11:694–700
20. Binsalamah ZM, Paul A, Khan AA, Prakash S, Shum-Tim D (2011) Intramyocardial sustained delivery of placental growth factor using nanoparticles as a vehicle for delivery in the rat infarct model. Int J Nanomedicine 6:2667–2678
21. Yang F, Cho SW, Son SM, Bogatyrev SR, Singh D, Green JJ, Mei Y, Park S, Bhang SH, Kim BS, Langer R, Anderson DG (2010) Genetic engineering of human stem cells for enhanced angiogenesis using biodegradable polymeric nanoparticles. Proc Natl Acad Sci U S A 107:3317–3322

Part II
Anti-angiogenesis Targets and Clinical Applications

Chapter 9
Survey of Anti-angiogenesis Strategies

Shaker A. Mousa

Abstract The field of angiogenesis modulation is at a major crossroad. Tremendous advances in basic science in this field are providing excellent support to the concept that is in contrast to a lack of strong clinical support at this point. With regard to the big gap between experimental data and clinical data, the best model of human malignancy is human malignancy, and the best models of human diabetic retinopathy (DR) and age-related macular degeneration (AMD) are in human DR and AMD patients. Clinical outcome should include benefit/risk ratio, hard end points (mortality and quality of life as opposed to increased microvascular density with pro-angiogenesis agents or tumor size reduction with anti-angiogenesis agents), and cost effectiveness.

Preclinical models should be used to provide guidance, placebo effect, comparative data, and mechanistic understanding as opposed to being used for expected clinical efficacy. We also have to understand existing strategy and how angiogenesis modulation can add further value in improving the efficacy and safety of new targets. The redundancy and multiplicity by which blood vessels are remodeled might account for the acquired resistance observed with single mechanism-based anti-angiogenesis strategies. Hence, improving the efficacy would require an anti-angiogenesis target that works at key upstream or common downstream signaling switches that regulate angiogenesis at multiple levels.

The anti-VEGF agent bevacizumab was approved as anti-angiogenesis therapy in 2004 for the treatment of colon cancer and other tumor types. Since then, an array of anti-angiogenesis agents have been developed and approved for the treatment of cancer in conjunction with chemotherapy or as monotherapy in age-related macular degeneration (AMD). Single mechanism-based anti-angiogenesis therapies for

S.A. Mousa (✉)
The Pharmaceutical Research Institute at Albany College of Pharmacy
and Health Sciences, Rensselaer, NY, USA
e-mail: shaker.mousa@acphs.edu

S.A. Mousa and P.J. Davis (eds.), *Angiogenesis Modulations in Health and Disease:*   95
Practical Applications of Pro- and Anti-angiogenesis Targets,
DOI 10.1007/978-94-007-6467-5_9, © Springer Science+Business Media Dordrecht 2013

cancer are facing a number of hurdles in terms of the level of efficacy, suggesting a need for more effective, broad spectrum anti-angiogenesis agents that work at key upstream pathways. Hence, achieving efficient anti-angiogenesis response will require approaches to simultaneously or sequentially targeting multiple aspects of the tumor microenvironment.

Background

In normal physiological angiogenesis, new blood vessels are formed at a slow rate during normal tissue growth, repair, and fetus development during pregnancy. In pathological angiogenesis, tumors cannot grow or metastasize without the generation of new blood vessels from existing ones that allow for the supply of oxygen and nutrients [1].

The key cells for new blood vessel formation are the endothelial cells along with the support of other cells such as pericyte and smooth muscle cells. For new blood vessel formation, existing vascular endothelial cells get activated and secrete matrix degrading enzymes that degrade the extracellular matrix (ECM) to allow for its migration and invasion, which allows the assembly of those endothelial cells to then assemble a new network of tube-like structures, forming new blood vessels with lumen for blood perfusion [1–3]. The balance between pro- and anti-angiogenesis factors allows for the hemostasis and the regulation of new blood vessel growth [1, 2]. See Table 9.1 for a list of anti-angiogenesis mediators.

Table 9.1 Anti-angiogenesis factors/mediators

Anti-angiogenesis factors/mediators
Angiostatin (Plasminogen fragment)
Endostatin (Collagen XVIII fragment)
Plasminogen Activator Inhibitor (PAI-1)
Vasostatin
Prolactin
Thrombospondin-1
Platelet factor 4
Interferon-γ (IF-γ)
Kininogen domain 5
Glycosaminoglycan (heparins)
Anti-VEGF
Thyroid antagonists (tetrac, triac)
Nicotinic receptor antagonists
Cathepsin inhibitors
Histone deacylase inhibitors
αv integrin antagonists
β1 integrin antagonists
Anti-tissue factor/Anti-factor VIIa (TFPI)
Antioxidants
TIMP

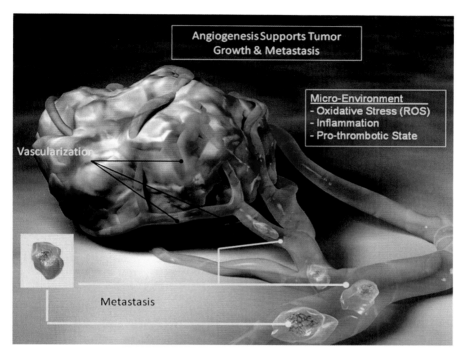

Fig. 9.1 Angiogenesis in tumor growth and metastasis: impact of tumor microenvironment

There are a number of pro-angiogenesis factors that are known to activate endothelial cell tube formation and new blood vessel generation (Table 9.1) [4]. Similarly, there are a number of endogenous mediators that inhibit endothelial cell functions that lead to new blood vessel formation (Table 9.1) [1, 2, 5, 6]. The incidence of tumor metastasis is associated with accelerated pathological angiogenesis and poor clinical outcome (Fig. 9.1) [1, 3].

Continued preclinical and clinical investigations identified major drawbacks (resistance to the treatment and increased thrombotic events) to the FDA-approved, first anti-angiogenesis anti-VEGF agent, namely bevacizumab, for treatment of metastatic colorectal cancer [7–11]. Several other anti-angiogenesis agents are either approved or in different stages of clinical trials (Table 9.2). In addition, lack of thoroughly validated predictive biomarkers has been one of the major hurdles to stratify cancer patients and to monitor tumor progression and response to the therapy. Investigations in clinic and preclinical settings have provided some insight for these challenges. There are mechanisms by which dynamic changes occur in the tumor microenvironment in response to anti-angiogenesis therapy, leading to drug resistance. These mechanisms include direct selection of clonal cell populations with the capacity to rapidly up-regulate alternative pro-angiogenesis pathways, increased invasive capacity, and intrinsic resistance to hypoxia.

Table 9.2 Angiogenesis inhibitors approved and under clinical development

Drug	Status	Target/mechanism	Indication
Bevacizumab (Avastin®)	FDA-approved[a]	Anti-VEGF mAb	Breast, colorectal, bladder, and other cancer
Ranibizumab (Lucentis®)	FDA-approved Phase III	Anti-VEGF (VEGR-2)	AMD HCC, Melanoma, NSCLC, RCC
Pegaptanib (Macugen®)	FDA-approved	VEGF inhibitor	AMD
Bortezomib (Velcade®)	FDA-approved	Reversible inhibitor of the 26S proteasome	Multiple myeloma
Thalidomide (Thalomid®)	FDA-approved	Suppression of TNF-α production and down-regulation of adhesion molecules	Multiple myeloma
Lenalidomide (Revlimid®)	FDA-approved	Analogue of thalido-mide – complex actions	Myelodysplastic syndromes and multiple myeloma
Erlotinib (Tarceva®)	FDA-approved	EGFR inhibitor	NSCLC, pancreatic cancers
Sorafenib (Nexavar®)	FDA-approved	VEGFR and PDGFR inhibitor	HCC and RCC
Sunitinib (Sutent®)	FDA-approved	VEGFR and PDGFR inhibitor –Multikinase	RCC and GI cancer
Aflibercept (Eylea®)	FDA-approved Phase III	VEGF trap	AMD Colorectal cancer
MEDI 522 (Vitaxin®)	Phase II	Anti-$\alpha v \beta 3$ mAb	Metastatic cancer
Cilengitide	Phase II/III	Cyclic peptide, inhibit $\alpha v \beta 3 / \beta 5$	GBM
Mecamylamine (ATG003)	Phase II	Non specific AChR noncompetitive antagonist	AMD and diabetic macular edema
SU6668	Phase I	Inhibit VEGFR, PDGFR, and FGFRs	Advanced solid tumors
Vandetanib (Zactima®)	Phase I	VEGFR and PDGFR inhibitor	NSCLC, thyroid cancers

To date the FDA has approved the use of several agents as angiogenesis inhibitors in the treatment of neoplastic and non-neoplastic diseases, whereas other drugs are still under clinical trials. Data were taken from www.cancer.gov, www.fda.gov, www.clinicaltrials.gov

Several other biologic agents and small molecules not listed are in different stages of clinical development

mAb monoclonal antibody, *AMD* age-related macular degeneration, *HCC* hepatic cell carcinoma, *RCC* renal cell carcinoma, *GI* gastrointestinal, *NSCLC* non-small cell lung cancer, *GBM* glioblastoma

[a]Avastin has been withdrawn in the US but is available in the EU and Japan for breast cancer

Strategies

There are two key strategies for the inhibition of angiogenesis:

1. Interfere with pro-angiogenesis factor(s) using either an antibody that binds with high affinity to that factor or a receptor antagonist that blocks its action at the receptor level.
2. Inhibit at different levels the process of endothelial tube formation and assembly of blood vessels (migration, proliferation, ECM degradation, and invasion).

The most advanced anti-angiogenesis targets might not be optimal, and it might require combinations with cytotoxic agent(s) such as is the case in cancer management. The αv integrin antagonist cyclic peptide cilengitide targets αv integrin in both cancer cells and vascular endothelium [12] and is currently in phase III clinical trial for glioblastoma patients [13]. Additionally, cilengitide is in phase II non-metastatic castration resistant prostate cancer trials [14].

Angiogenesis and Angiogenesis Inhibitors: A New Potential Anti-cancer Therapeutic Strategy

The angiogenesis process is a complex multi-step cascade under the control of positive and negative regulators along with other factors in the microenvironment [1, 3].

Biotechnology-driven Strategies

Kininostatin

Domain5 (D5) of high molecular weight kininogen has been shown to inhibit endothelial cell migration toward vitronectin, endothelial cell proliferation and angiogenesis in the chicken chorioallantoic membrane assay. Structure-function studies demonstrate that D5 contains a peptide that only inhibits migration and another peptide that inhibits only proliferation. Data on a related inhibitory monoclonal antibody confirmed the role of kininiogen in angiogenesis modulation [15].

The α3 Chain of Type IV Collagen Inhibits Tumor Growth

Studies from our laboratory have shown that type IV collagen (COL IV) and specifically a peptide comprising residues 185–203 of the non-collagenous domain (NCI) of the α3 (IV) chain has several biological activities including: inhibition of neutrophil activation, promotion of tumor cell adhesion and chemotaxis, and inhibition of tumor cell proliferation [16].

Anti-VEGF

VEGF is an endothelial cell-specific mitogen and an angiogenic inducer. VEGF is essential for developmental angiogenesis and is also required for female reproductive functions and bone formation. Substantial evidence also implicates VEGF in tumors and intraocular neovascular syndromes. Currently, a humanized anti-VEGF monoclonal antibody (rhuMab VEGF) is in clinical trials as a treatment for various solid tumors. Phase II studies in patients with colorectal and lung cancer, where anti-VEGF was used in conjunction with standard chemotherapy, have shown evidence of clinical efficacy [17, 18]. Additionally, various anti-VEGF strategies demonstrated successful outcome in various ocular neovascularization disorders [19, 20]. In addition to antagonizing VEGF, there are also small molecules that inhibit receptor tyrosine kinases (RTKs). Many RTK inhibitors have been described that exhibit different specificity profiles.

Role of Matrix Metalloproteinase in Cancer

Matrix Metalloproteinase (MMPs) are zinc dependent endopeptidases that degrade extracellular matrix proteins (collagens, laminin, and fibronectin) during the process of cancer invasion and metastasis. Contrary to expectations, MMPs are produced primarily by reactive stromal and inflammatory cells surrounding tumors rather than cancer cells. Potent, orally available, hydroxamic acid-derived inhibitors of MMPs were in clinical trials for the treatment of lung, prostate, pancreatic, gastric, and breast cancer but safety profiles along with limited efficacy halted this class [21, 22].

2-Methoxyestradiol

2-Methoxyestradiol (2ME) is now being recognized as a potent, orally active anti-angiogenesis and antitumor agent that has minimal if any toxicity. Despite being an endogenous estradiol metabolite, 2ME does not show any estrogenic activity *in vitro* and *in vivo* at doses at which estradiol and other metabolites are clearly estrogenic. 2ME is presently in a phase II trial for metastatic cancer alone and in combination with sunitinib [23].

Small Molecule Integrin Antagonists

The role of various integrins in angiogenesis-mediated disorders was demonstrated [24]. In addition to the well-described role of αvß3 and αvß5 integrins [25], a role for α5ß1 integrin in the modulation of angiogenesis was defined [26]. The role of α5ß1 integrin in angiogenesis was established through the use of monoclonal antibody, cyclic peptide, and non-peptide α5ß1 integrin antagonists [26].

Nutraceutical-derived Polyphenol Angiogenesis Inhibitors

There are several evidences for the potential beneficial effect of phytochemicals on cancer-related pathways, particularly with regard to anti-angiogenesis. Plant phenolics are the most important category of phytochemicals, including flavonoids. Prominent phytochemicals affecting different pathways of angiogenesis are green tea polyphenols (epigallocatechin gallate, EGCG), soy bean isoflavones (genistein), and several other polyphenols derived from natural sources [27–29].

Anti-angiogenesis Therapy in Multiple Myeloma

Thalidomide that was withdrawn in Europe earlier because of its teratogenicity in pregnant mothers when used as a sedative, demonstrated potent inhibition of angiogenesis [30]. Thalidomide and analogues were demonstrated to be effective in relapsed/refractory multiple myeloma with mild systemic toxicity. Encouraging synergistic outcome was demonstrated in studies of thalidomide when combined with dexamethasone or conventional chemotherapy [30]. Other studies are examining the potential of thalidomide analogues, inhibitors of VEGF isoforms, VEGF receptors, and endostatin [30, 31]. However, those angiogenesis inhibitors such as thalidomide and the anti-VEGF (SU015) demonstrated increased incidence of thrombosis when used in combination with chemotherapeutic agents [32, 33]. Bortezomib during induction and maintenance improves complete response and achieves superior progression-free survival and overall survival [34].

The Spanish Myeloma Group conducted a trial to compare bortezomib/thalidomide/dexamethasone (VTD) versus thalidomide/dexamethasone (TD) versus vincristine, BCNU, melphalan, cyclophosphamide, prednisone/vincristine, BCNU, doxorubicin, dexamethasone/bortezomib (VBMCP/VBAD/B) in patients aged 65 years or younger with multiple myeloma. The primary endpoint was complete response rate post-induction and post-autologous stem cell transplantation (ASCT). Results showed that VTD is a highly effective induction regimen prior to ASCT [35].

Anticoagulants and Angiogenesis

Many cancer patients reportedly have a hyper-coaguable state, with recurrent thrombosis due to the impact of cancer cells and chemotherapy on the coagulation cascade [15]. Analysis of biomarkers of the coagulation cascade and of vessel wall activation was performed and showed significant increases in thrombin generation and endothelial cell perturbation in a treatment cycle-dependent manner when combining angiogenesis inhibitors and chemotherapeutic agents [32, 33]. The incidence of thromboembolic events, possibly related to the particular regimen tested in this study (SU015 and chemotherapeutic agents), discourages further investigation of this regimen [33]. This investigation along with the increased incidences of deep vein thrombosis (DVT) in multiple myeloma patients receiving thalidomide and chemotherapeutic agents [32] suggest the potential advantages of using an anticoagulant such as heparin or low molecular weight heparin (LMWH). Additionally, studies have demonstrated that unfractionated heparin (UFH) or LMWH interfere with various processes involved in tumor growth and metastasis. Clinical trials have indicated a clinically relevant effect of LMWH, as compared to UFH, on the survival of cancer patients with DVT. Mechanism and efficacy of the LMWH and its *in vivo* releasable TFPI on the regulation of angiogenesis and tumor growth was documented [36, 37]. Heparin, steroids, and heparin/steroid combinations have been used in a variety of *in vitro* models and *in vivo* in animal models as effective inhibitors of angiogenesis [38, 39]. Additionally, platelet-tumor cell interactions could play a significant role in tumor metastasis [39].

In the field of dermatology, a number of FDA-approved agents have anti-angiogenesis properties:

(a) Alitretinoin (Panretin® 0.1 % gel, Ligand) is a topical retinoid (vitamin A) indicated for the treatment of AIDS-related Kaposi's sarcoma.
(b) Imiquimod (Aldara® 5 % cream, Zyclara 3.75 % cream, Graceway) is a Toll-Like Receptor 7 agonist that might exert anti-angiogenesis activity through down-regulation of FGF-2 and MMP-9, and induction of endothelial apoptosis. Imiquimod is indicated for both benign neoplasms (genital warts) and for malignant skin cancers (actinic keratosis and basal cell carcinoma).
(c) Polyphenon E (Veregen® 15 % ointment, Bradley/MediGene) is a defined composition of polyphenolic EGCG extracted from green tea leaves, which inhibit angiogenesis. Polyphenon E topical ointment is indicated for genital warts.

Critical Issues

The potential development of resistance would exist specially in the case of single mechanism-based anti-angiogenesis strategies because of the redundancy in the pro-angiogenesis pathways [40]. Hence, the need for a safe and broad spectrum anti-angiogenesis agent still exists versus the combination of multiple single mechanism-based anti-angiogenesis agents. An anti-angiogenesis agent is not meant to be a monotherapy but rather meant to be combined with chemotherapeutic agent(s) at the same time or in different sequences for optimal efficacy.

Anti-angiogenesis Agents and Thrombosis

Anti-angiogenesis agents might mediate endothelial cell dysfunction, which could be associated with increased incidence of thrombosis [31–33, 41].

Diagnostic Imaging of Angiogenesis

Angiogenesis is a novel field that promises to provide new venues for blood flow in patients with severe ischemic heart disease and severe peripheral vascular disease, and it promises to control malignancies by controlling blood supply. However, it became clear early on that a novel field necessitates novel outcome measures that follow its beneficial effects. There is a growing need for non invasive methods to serially assess the status of the coronary, peripheral, and tumor vasculature. These might include magnetic resonance imaging and monitoring of angiogenesis, nuclear perfusion imaging of angiogenesis, and other modalities. Biomarkers might reflect the prognosis of the diseases as well as the effectiveness of the treatment [42].

Micro RNA and Angiogenesis Modulation

Recent reports have shown that the expression pattern of micro RNA (miRNA) levels is a common feature in many human diseases including those involving the vasculature such as cardiovascular disease, ocular disease, and cancer [43]. This has opened a novel strategy for targeted pro- or anti-angiogenesis therapies since a single miRNA can potentially regulate endothelial functions by targeting multiple transcription pathways. The miRNA strategy can be utilized to either promote angiogenesis by targeting negative regulators in angiogenesis signaling pathways or inhibit angiogenesis by targeting positive regulators. Nanotechnology and targeted delivery of miRNA or anti-miRNA could be delivered to tumor endothelium using targeted nanoparticles [44]. Lessons learned from antisense technologies and RNA interference approaches will no doubt be relevant in advancement of miRNA therapeutics in addition to the use of nano targeted delivery of mRNA or anti-mRNA. However, a potential limitation of miRNA-based therapy is the possible off-target effects that might lead to serious adverse effects [45, 46].

References

1. Mousa SA (2000) Mechanisms of angiogenesis in vascular disorders: potential therapeutic targets. In: Mousa SA (ed) Angiogenesis inhibitors & stimulators: potential therapeutic implications. Landes Bioscience (Autsin, TX), pp 1–12, Chapter 1
2. Pavlakovic H, Havers W, Schweigerer L (2001) Multiple angiogenesis stimulators in a single malignancy: implications for anti-angiogenic tumor therapy. Angiogenesis 4(4):259–262
3. Ranieri G, Gasparini G (2001) Angiogenesis and angiogenesis inhibitors: a new potential anti-cancer therapeutic strategy. Curr Drug Targets Immune Endocrinol Metab Disord 1(3):241–253
4. Carmeliet P, Jain RK (2011) Molecular mechanisms and clinical applications of angiogenesis. Nature 473(7347):298–307
5. Ali SH, O'Donnell AL, Balu D, Pohl MB, Seyler MJ, Mohamed S, Mousa S, Dandona P (2000) Estrogen receptor-α in the inhibition of cancer growth and angiogenesis. Cancer Res 60:7094–7098
6. Kerbel RS (2008) Tumor angiogenesis. N Engl J Med 358(19):2039–2049
7. Chamberlain MC (2011) Bevacizumab for the treatment of recurrent glioblastoma. Clin Med Insight Oncol 5:117–129
8. Bottsford-Miller JN, Coleman RL, Sood AK (2012) Resistance and escape from antiangiogenesis therapy: clinical implications and future strategies. J Clin Oncol 30(32):4026–4034
9. Rapisarda A, Melillo G (2012) Overcoming disappointing results with antiangiogenic therapy by targeting hypoxia. Nat Rev Clin Oncol 9(7):378–390
10. Bergers G, Hanahan D (2008) Modes of resistance to anti-angiogenic therapy. Nat Rev Cancer 8(8):592–603
11. Belcik JT, Qi Y, Kaufmann BA, Bullens S, Morgan TK, Bagby SP, Kolumam G, Kowalski J, Oyer JA, Bunting S, Lindner JR (2012) Cardiovascular and systemic microvascular effects of anti-vascular endothelial growth factor therapy for cancer. J Am Coll Cardiol 60(7):618–625
12. Kurozumi K, Ichikawa T, Onishi M, Fujii K, Date I (2012) Cilengitide treatment for malignant glioma: current status and future direction. Neurol Med Chir (Tokyo) 52(8):539–547
13. Scaringi C, Minniti G, Caporello P, Enrici RM (2012) Integrin inhibitor cilengitide for the treatment of glioblastoma: a brief overview of current clinical results. Anticancer Res 32(10):4213–4223
14. Alva A, Slovin S, Daignault S, Carducci M, Dipaola R, Pienta K, Agus D, Cooney K, Chen A, Smith DC, Hussain M (2012) Phase II study of cilengitide (EMD 121974, NSC 707544) in patients with non-metastatic castration resistant prostate cancer, NCI-6735. A study by the DOD/PCF prostate cancer clinical trials consortium. Invest New Drugs 30(2):749–757
15. Colman RW, Jameson BA, Lin Y, Mousa S (2000) Inhibition of angiogenesis by kininogen domain 5. Blood 95(2):543–550
16. Shahan TA, Grant DS, Tootell M, Ziaie Z, Ohno N, Mousa S, Mohamad S, Delisser H, Kefalides N (2004) Oncothanin, a peptide from alpha3 chain of type IV collagen, inhibits tumor growth by inhibiting angiogenesis. Connect Tissue Res 45(3):151–163
17. Small AC, Oh WK (2012) Bevacizumab treatment of prostate cancer. Expert Opin Biol Ther 12(9):1241–1249
18. de Groot JF, Lamborn KR, Chang SM, Gilbert MR, Cloughesy TF, Aldape K, Yao J, Jackson EF, Lieberman F, Robins HI, Mehta MP, Lassman AB, Deangelis LM, Yung WK, Chen A, Prados MD, Wen PY (2011) Phase II study of aflibercept in recurrent malignant glioma: a North American Brain Tumor Consortium study. J Clin Oncol 29(19):2689–2695
19. Mitchell P (2011) A systematic review of the efficacy and safety outcomes of anti-VEGF agents used for treating neovascular age-related macular degeneration: comparison of ranibizumab and bevacizumab. Curr Med Res Opin 27(7):1465–1475
20. Chong V (2012) Biological, preclinical and clinical characteristics of inhibitors of vascular endothelial growth factors. Ophthalmologica 227(Suppl 1):2–10
21. Moore MJ, Hamm J, Dancey J, Eisenberg PD, Dagenais M, Fields A, Hagan K, Greenberg B, Colwell B, Zee B, Tu D, Ottaway J, Humphrey R, Seymour L (2003) National Cancer

Institute of Canada Clinical Trials Group. Comparison of gemcitabine versus the matrix metalloproteinase inhibitor BAY 12–9566 in patients with advanced or metastatic adenocarcinoma of the pancreas: a phase III trial of the National Cancer Institute of Canada Clinical Trials Group. J Clin Oncol 21(17):3296–3302

22. Rudek MA, Fig. WD, Dyer V, Dahut W, Turner ML, Steinberg SM, Liewehr DJ, Kohler DR, Pluda JM, Reed E (2001) Phase I clinical trial of oral COL-3, a matrix metalloproteinase inhibitor, in patients with refractory metastatic cancer. J Clin Oncol 19(2):584–592

23. Bruce JY, Eickhoff J, Pili R, Logan T, Carducci M, Arnott J, Treston A, Wilding G, Liu G (2012) A phase II study of 2-methoxyestradiol nanocrystal colloidal dispersion alone and in combination with sunitinib malate in patients with metastatic renal cell carcinoma progressing on sunitinib malate. Invest New Drugs 30(2):794–802

24. Mousa SA (2001) IBC's 6th Annual conference on angiogenesis: novel therapeutic developments. Expert Opin Investig Drugs 10(2):387–391

25. Van Waes C, Enamorado I, Hecht D, Sulica L, Chen Z, Batt G, Mousa S (2000) Effects of the novel alpha v integrin antagonist SM256 and cis-platinum on growth of murine squamous cell carcinoma PAMLY8. Int J Oncol 16(6):1189–1195

26. Kim S, Mousa S, Varner J (2000) Requirement of integrin alpha 5 beat 1 and its ligand fibronectin in angiogenesis. Am J Pathol 156:1345–1362

27. Siddiqui IA, Adhami VM, Bharali DJ, Hafeez BB, Asim M, Khwaja SI, Ahmad N, Cui H, Mousa SA, Mukhtar H (2009) Introducing nanochemoprevention as a novel approach for cancer control: proof of principle with green tea polyphenol epigallocatechin-3-gallate. Cancer Res 69(5):1712–1716

28. Nguyen MM, Ahmann FR, Nagle RB, Hsu CH, Tangrea JA, Parnes HL, Sokoloff MH, Gretzer MB, Chow HH (2012) Randomized, double-blind, placebo-controlled trial of polyphenon E in prostate cancer patients before prostatectomy: evaluation of potential chemopreventive activities. Cancer Prev Res (Phila) 5(2):290–298

29. Kanai M, Yoshimura K, Asada M, Imaizumi A, Suzuki C, Matsumoto S, Nishimura T, Mori Y, Masui T, Kawaguchi Y, Yanagihara K, Yazumi S, Chiba T, Guha S, Aggarwal BB (2011) A phase I/II study of gemcitabine-based chemotherapy plus curcumin for patients with gemcitabine-resistant pancreatic cancer. Cancer Chemother Pharmacol 68(1):157–164

30. Brower V (2012) Lenalidomide maintenance for multiple myeloma. Lancet Oncol 13(6):e238

31. Tosi P, Tura S (2001) Antiangiogenic therapy in multiple myeloma. Acta Haematol 106(4):208–213

32. Bennett CL, Schumock GT, Desai AA, Kwaan HC, Raisch DW, Newlin R, Stadler W (2002) Thalidomide-associated deep vein thrombosis and pulmonary embolism. Am J Med 113(7):603–606

33. Kuenen BC, Rosen L, Smit EF, Parson MR, Levi M, Ruijter R, Huisman H, Kedde MA, Noordhuis P, van der Vijgh WJ, Peters GJ, Cropp GF, Scigalla P, Hoekman K, Pinedo HM, Giaccone G (2002) Dose-finding and pharmacokinetic study of cisplatin, gemcitabine, and SU5416 in patients with solid tumors. J Clin Oncol 20(6):1657–1667

34. Sonneveld P, Schmidt-Wolf IG, van der Holt B et al (2012) Bortezomib induction and maintenance treatment in patients with newly diagnosed multiple myeloma: results of the randomized phase III HOVON-65/GMMG-HD4 trial. J Clin Oncol 30(24):2946–2955

35. Rosiñol L, Oriol A, Teruel AI, Hernández D, (PETHEMA/GEM) group et al (2012) Superiority of bortezomib, thalidomide, and dexamethasone (VTD) as induction pretransplantation therapy in multiple myeloma: a randomized phase 3 PETHEMA/GEM study. Blood 120(8):1589–1596

36. Mousa S (2002) Angiogenesis, coagulation activation, and malignant dissemination. Semin Thromb Hemost 28(1):45–52

37. Mousa SA, Mohamed S (2004) Anti-angiogenic mechanisms and efficacy of the low molecular weight heparin, tinzaparin: anti-cancer efficacy. Oncol Rep 12(4):683–688

38. Jung SP, Siegrist B, Wade MR, Anthony CT, Woltering EA (2001) Inhibition of human angiogenesis with heparin and hydrocortisone. Angiogenesis 4(3):175–186

39. McCarty OJT, Mousa S, Bray P, Konstantopoulos K (2000) Immobilized platelet support human colon carcinoma cell tethering, rolling and firm adhesion under dynamic flow conditions. Blood 96(5):1789–1797
40. Kerbel RS, Yu J, Tran J, Man S, Viloria-Petit A, Klement G, Coomber BL, Rak J (2001) Possible mechanisms of acquired resistance to anti-angiogenic drugs: implications for the use of combination therapy approaches. Cancer Metastasis Rev 20(1–2):79–86
41. Kaushal V, Kohli M, Zangari M, Fink L, Mehta P (2002) Endothelial dysfunction in antiangiogenesis-associated thrombosis. J Clin Oncol 20(13):3042–3043
42. Chen Z, Malhotra P, Thomas G, Ondrey F, Duffey D, Smith C, Enamorada D, Yeh N, Kroog G, Rudy S, McCullagh L, Mousa S (1999) Expression of proinflammatory and proangiogenic cytokines in patients with head and neck cancer. Clin Cancer Res 5:1369–1379
43. Liu N, Olson EN (2010) MicroRNA regulatory networks in cardiovascular development. Dev Cell 18:510–525
44. Murphy EA, Majeti BK, Barnes LA, Makale M, Weis SM, Lutu-Fuga K, Wrasidlo W, Cheresh DA (2008) Nanoparticle-mediated drug delivery to tumor vasculature suppresses metastasis. Proc Natl Acad Sci U S A 105(27):9343–9348
45. Heusschen R, van Gink M, Griffioen AW, Thijssen VL (2010) MicroRNAs in the tumor endothelium: novel controls on the angioregulatory switchboard. Biochim Biophys Acta 1805:87–96
46. Rayner KJ, Suarez Y, Davalos A, Parathath S, Fitzgerald ML, Tamehiro N, Fisher EA, Moore KJ, Fernandez-Hernando C (2010) MiR-33 contributes to the regulation of cholesterol homeostasis. Science 328:1570–1573

Chapter 10
Tetraiodothyroacetic Acid (Tetrac), Nanotetrac and Anti-angiogenesis

Paul J. Davis, Faith B. Davis, Mary K. Luidens, Hung-Yun Lin, and Shaker A. Mousa

Abstract Tetraiodothyroacetic acid (tetrac) is a naturally occurring derivative of thyroid hormone, T_4. In the absence or presence of L-T4 or L-T3, tetrac has been found to disrupt a number of functions or events that are important to cancer cells via the known thyroid hormone-tetrac receptor on the plasma membrane integrin $\alpha v \beta 3$. These functions include regulation of cell division, local stimulation of angiogenesis, chemo-resistance and resistance to radiation. It is desirable to reformulate tetrac as a nanoparticle whose activity is exclusively at the cell surface integrin. Nanotetrac has been designed to limit tetrac to the extracellular space on the basis of the size of the nanoparticle and to provide optimized exposure of the biphenyl structure and acetic acid side chain of its inner ring to the receptor site on $\alpha v \beta 3$. Tetrac and its novel nanoparticulate formulation have anti-angiogenesis activity that transcends the inhibition of thyroid hormone-binding at the integrin. Restriction of nanotetrac to the extracellular space has been verified, and nanotetrac has been shown to be up to 10-fold more potent than unmodified tetrac at its integrin target. Nanotetrac formulations have potential applications as inhibitors of tumor-related angiogenesis

P.J. Davis (✉)
The Pharmaceutical Research Institute at Albany College of Pharmacy and Health Sciences, Rensselaer, NY, USA

Department of Medicine, Albany Medical Center, Albany, NY, USA
e-mail: pdavis.ordwayst@gmail.com

F.B. Davis • S.A. Mousa
The Pharmaceutical Research Institute at Albany College of Pharmacy and Health Sciences, Rensselaer, NY, USA
e-mail: shaker.mousa@acphs.edu

M.K. Luidens
Department of Medicine, Albany Medical College, Albany, NY, USA

H.-Y. Lin
Institute of Cancer Biology and Drug Discovery, Taipei Medical University, Taipei, Taiwan

S.A. Mousa and P.J. Davis (eds.), *Angiogenesis Modulations in Health and Disease: Practical Applications of Pro- and Anti-angiogenesis Targets*, DOI 10.1007/978-94-007-6467-5_10, © Springer Science+Business Media Dordrecht 2013 and of angiogenesis that is unrelated to malignancy, including clinically significant disorders ranging from skin diseases to vascular proliferation in the retina and neovascularization associated with inflammatory states.

Biochemical History of Tetrac

Tetraiodothyroacetic acid (tetrac) is a naturally occurring derivative of thyroid hormone (L-thyroxine, T_4). It accounts for less than 1 % of circulating thyroid hormone. Inside human cells, tetrac is a low-grade thyromimetic, that is, it has low-potency actions that resemble those of 3, 5, 3'-triiodo-L-thyronine (T_3), the most active form of thyroid hormone [1]. For 30 or more years, tetrac has been known to be taken up by pituitary cells that secrete thyrotropin (TSH) and to inhibit endogenous human TSH release by the pituitary gland [2]. This action has been considered potentially useful in the clinic in the setting of TSH-dependent thyroid cancer.

Tetrac was also found to compete with T_4 for thyroid hormone-binding sites on human serum pre-albumin (transthyretin, TTR) [3], but not for sites on human thyroxine-binding globulin (TBG), the principal transport protein for iodothyronines in human serum. These observations caused us in the 1980s to test tetrac for its ability to block the nongenomic actions of thyroid hormone that we demonstrated to exist in human mature red blood cell (RBC) membranes [4, 5] and intracellular membranes, such as those of the sarcoplasmic reticulum [6]. These actions included Ca^{2+} transport by calmodulin-responsive Ca^{2+}-ATPase [7]. Such effects were 'nongenomic' because they were independent of the nuclear thyroid hormone receptor (TR) and gene transcription [8]. Tetrac indeed inhibited such actions of T_4 and T_3 and thus became a probe for certain actions of thyroid hormone that were initiated at the plasma membrane, although the cell surface receptor site for the hormone was not identified until 2005 [9].

Studies focused on the plasma membrane receptor for thyroid hormone on integrin $\alpha v\beta3$ revealed that this receptor mediated the pro-angiogenic action of T_4 and T_3, and that tetrac was anti-angiogenic in terms of its ability to block this action [10]. As will be discussed below, tetrac and a novel nanoparticulate formulation of the agent have anti-angiogenic activity that transcends the inhibition of thyroid hormone-binding at the integrin. Tetrac formulations also have anti-proliferative activity against tumor cells and have been shown to have chemosensitizing and radiosensitizing effects in cancer cells [10, 11].

Integrin $\alpha v\beta3$ Contains a Thyroid Hormone-Tetrac Receptor

J.J. Bergh et al. in 2005 [9] described the existence of a thyroid hormone-tetrac receptor on plasma membrane integrin $\alpha v\beta3$. This heterodimeric integrin is generously expressed on rapidly dividing blood vessel cells and by cancer cells, enabling scanning technology focused on the integrin to detect tumors [12]. The fit of unmodified tetrac into the hormone-binding groove in the extracellular domain [11, 13] of the integrin permits the agent to block the binding of T_4 and T_3 and inhibit actions of these agonist forms of thyroid hormone on cancer cell proliferation and cancer-related angiogenesis as discussed in Chap. 4. The integrin is found on virtually all cancer cells and reads the presence of specific extracellular matrix proteins that are relevant to tumor cell migration and tumor mass formation.

In the absence or presence of T_4 and T_3, however, tetrac has been found to disrupt via $\alpha v \beta 3$ a number of other functions or events that are important to cancer cells. These functions include regulation of cell division, local stimulation of angiogenesis, chemoresistance and resistance to radiation [10, 14]. It was surprising to find that, acting at the cell surface, tetrac coherently interfered with expression of differentially regulated genes whose products included the cyclins and *thrombospondin 1* [15]. The *thrombospondin 1* gene is usually silent in tumor cells because it suppresses angiogenesis. Tetrac also blocked the actions of vascular endothelial growth factor (VEGF) and basic fibroblast growth factor (bFGF) [16], released by cancer cells to promote angiogenesis in an autocrine manner. The mechanism of tetrac involved here is thought to involve disorganization of crosstalk between the integrin and nearby receptors for VEGF [17] and bFGF [10, 18] and inhibition of local release of bFGF [18]. The hormone analogue was found to cause retention by cancer cells of traditional chemotherapeutic agents to which the cells previously showed resistance [19]. These agents included doxorubicin, etoposide, and cisplatin. Tetrac was also shown to inhibit the ability of cancer cells to repair double-strand DNA breaks that radiation induces [20, 21], thus radiosensitizing these cells. Finally, expression of genes for cytokines involved in inflammation and relevant to inflammation-associated cancer was shown to be blocked by tetrac [15].

The foregoing describes the anti-cancer and anti-angiogenic features of tetrac manifested at integrin $\alpha v \beta 3$. However, unmodified tetrac is taken up by cells and has, within the cell, low-potency thyromimetic activity [1, 2]. This can include proliferative—rather than anti-proliferative—behavior (H.Y. Lin: unpublished). While the algebraic sum of these anti-proliferative and proliferative effects favors anti-cancer activity, it is desirable to reformulate tetrac as a nanoparticle whose activity is exclusively at the integrin (Section "Nanoparticulate Tetrac (Nanotetrac)", below).

Nanoparticulate Tetrac (Nanotetrac)

S.A. Mousa and co-workers have constructed an approximately 250 nm poly(lactic-co-glycolic acid) (PLGA) formulation in which a limited number of tetrac moieties are covalently bound to and protrude from a PLGA nanoparticle [22, 23] (Fig. 10.1). Binding of tetrac to the PLGA is via an ether bond at the hydroxyl on the outer ring of tetrac (position 4) to a linker that is amide-bonded to the PLGA. Nanotetrac was

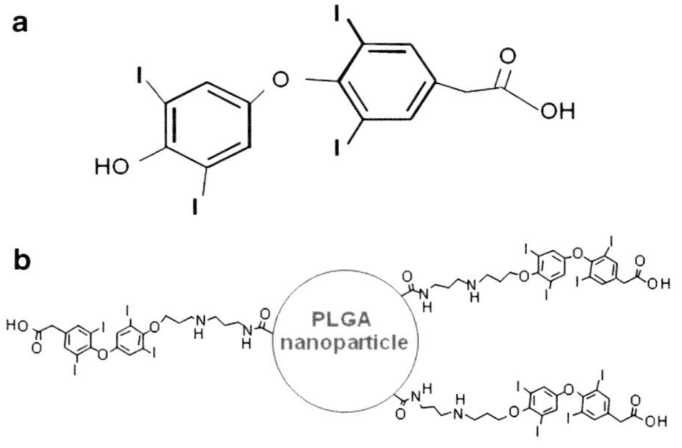

a

b

Fig. 10.1 Chemical structures of unmodified tetraiodothyroacetic acid (tetrac) (**a**) and nanoparticulate tetrac (nanotetrac) (**b**). An ether bond involving the outer ring hydroxyl group joins tetrac to a linker molecule which, in turn, is attached by an imbedded amide bond to the nanoparticle. Multiple tetrac moieties are bonded to the surface of the PLGA, enabling access of tetrac to its receptor groove in the extracellular domain of integrin $\alpha v \beta 3$

designed to limit tetrac to the extracellular space on the basis of the size of the nanoparticle and to provide optimized exposure of the biphenyl structure and acetic acid side chain of the inner ring to the receptor site on $\alpha v \beta 3$. Restriction of the molecule to the extracellular space has been verified [24].

Interestingly, the reformulation of tetrac as a nanoparticle has conferred two other qualities on the molecule. First, nanotetrac is up to 10-fold more potent than unmodified tetrac. The basis for this may reflect interaction(s) of the polymer (PLGA) chain, away from the tetrac moieties, with the 'legs' of the extracellular domain of $\alpha v \beta 3$, favoring 'on' kinetics. Another possibility is that the nanoparticulate tetrac ligand induces a long-lived change in the conformation of the $\alpha v \beta 3$ that persists after ligand dissociation and modulates activity of the integrin [25]. Second, and acting at the integrin receptor, nanotetrac induces a pattern of cancer survival gene expression that is somewhat different from that of unmodified tetrac [15]. There is 80–85 % congruence of gene expression caused by nanotetrac and tetrac. Examples of the disparities are down-regulation by nanotetrac in cancer cells of expression of the *epidermal growth factor receptor (EGFR)* gene, down-regulation of the majority of the members of the *Ras*-oncogene family, and up-regulation of the apoptosis inhibitor *MCL1* (*myeloid cell leukemia sequence 1*); these genes are unaffected by tetrac.

The difference in potency of tetrac and nanotetrac is readily seen in the comparison of efficacy of the compounds in human cancer xenografts in the nude mouse. Within 3 days of initiation of treatment, the two formulations cause a greater than 50 % decrease in the vascular supply of the xenografts, but this anti-angiogenic effect can be obtained with a dose of nanotetrac that is 10-fold less than that of tetrac.

Triiodothyroacetic Acid (Triac)

Triac (3, 3', 5-triiodothyroacetic acid), the deaminated derivative of T_3, can reproduce the inhibition of actions of T_4 and T_3 at the $\alpha v \beta 3$ thyroid hormone-tetrac receptor (H.Y. Lin: unpublished), but has not as yet been reformulated as a nanoparticle. These actions include inhibition of the pro-angiogenic activity of thyroid hormone studied in the chick chorioallantoic membrane (CAM) model. Inside the cell, triac is thyromimetic. It is thermogenic [26], has been shown in human subjects to have agonist thyroid hormone effects on liver and bone that are augmented compared to T_4 [27], and has been postulated to be a primordial form of thyroid hormone [28].

The unmodified compound is mentioned here for several reasons. Triac may be purchased over-the-counter in several European countries as a dietary supplement (tiratricol). The availability of this iodothyronine raises the possibility that anti-angiogenic side effects may be encountered in users of the agent. Such effects have not been reported, but it is unlikely that they have been anticipated, e.g., in the setting of wound-healing.

Potential Applications of Tetrac and Nanotetrac in Angiogenesis

Tetrac and Anti-angiogenesis in the Setting of Cancer

The clinical desirability of establishing control of angiogenesis in and about tumors is obvious and the feasibility of such control became apparent with the work of Judah Folkman and colleagues [29, 30]. In the context of cancers, local angiogenesis that supports tumor biology may be the result of more than a single vascular growth factor. These growth factors include VEGF, bFGF and other proteins [31]. Erythropoietin (EPO) may also be included among factors that enhance tumor-relevant angiogenesis [32].

In contrast to several of the anti-angiogenic agents used clinically (see below) that are designed to inhibit actions of single, specific vascular growth factors, tetrac and nanotetrac have been shown, in the absence or presence of agonist thyroid hormone (T_4 or T_3), to antagonize actions of multiple growth factors. These include VEGF, bFGF [17, 33], PDGF (S.A. Mousa: unpublished), EGF (S.A. Mousa: unpublished) and EPO [34]. This plural effect explains, at least in part, the rapid decrease in vascularity so far encountered in human tumor xenografts in mice exposed to systemic tetrac and nanotetrac [35–38] and the resultant decrease in volume of xenografts.

There are multiple mechanisms by which thyroid hormone analogues inhibit (nanotetrac or tetrac) or enhance (T_4 and T_3) the activities of various vascular growth factors. First, vascular growth factor gene expression in tumor cells or endothelial cells may be enhanced by thyroid hormone [18]. Second, iodothyronines may increase release of the growth factor(s) by the secreting cell, e.g., bFGF [18]. Third, crosstalk between the integrin and adjacent vascular growth receptors is well-described. The crosstalk may involve signal transducing biochemistry within or immediately below the plasma membrane. For example, inhibition by tetrac of mitogen-activated protein kinase (MAPK) activity alters activity of bFGF and other factors. Hormonal effects may in other cases depend upon interactions of the extracellular domains of the receptor(s) and the integrin [39] that could involve, in the case of VEGFR, the inhibition by tetrac formulations of dimerization of the growth factor receptor or obscuring of one or more of its immunoglobulin (Ig)-like domains [39]. Fourth, tetrac decreases abundance of *angiopoietin-2 (Ang-2)* mRNA in endothelial cells, but does not affect accumulation of *Ang-1* mRNA [17]. Ang-2 protein production in tumor vasculature anticipates or synergizes with vascular growth factor action to support tumor angiogenesis [40], whereas the Ang-1-Tie 2 system is a blood vessel-stabilizing pathway. This differential effect of tetrac with regard to angiogenesis is consistent with discrete, selective actions of tetrac or nanotetrac in cancer cells on elaboration of certain interleukins (see below) or of endogenous inhibitors or enhancers of apoptosis [15]. Fifth, tetrac can induce the expression of *thrombospondin 1*, an endogenous suppressor of angiogenesis that is almost invariably unexpressed in cancer cells [22]. Finally, T_4 and T_3 and tetrac may positively, in the case of the former, and negatively, in the case of tetrac, affect endothelial cell motility (S.A. Mousa: unpublished) that is important to neovascularization.

Currently available clinically are pharmaceuticals that affect single vascular growth factors. Bevacizumab (Avastin®) and ranibizumab (Lucentis®) are monoclonal antibodies to VEGF, developed as anti-angiogenic agents. Administered parenterally—intravenously, or, in the case of eye disease, intra-ocularly—these humanized antibodies are unquestionably effective in clinical disease settings in which VEGF or VEGF-A are contributory pathophysiologic factors. There are several subtypes of VEGF; VEGF-A [41], for example, is a form frequently released locally by tumor cells and which induces a porous vasculature. It is clear that the application of bevacizumab to the cancers for which it is approved is primarily adjunctive [42, 43] and is not curative. It has not so far been practical to produce for clinical use multi-monoclonal antibody preparations that are directed at more than a single vascular growth factor.

Bevacizumab is applied with U.S. Food and Drug Administration (FDA) approval to management of several forms of cancer and used without FDA approval in settings of unwanted angiogenesis in the absence of cancer, e.g., diabetic retinopathy [44]. Ranibizumab is an anti-angiogenic drug marketed for management of a form of retinal macular degeneration that may lead to loss of vision [45]. Bevacizumab and ranibizumab are discussed in more detail in Chaps. 12 and 13.

Application of Tetrac/Nanotetrac to Clinical Conditions of Excessive Angiogenesis Not Associated with Malignancy

Skin Disorders

Skin redness (erythema) in specific settings such as acne rosacea [46] or psoriasis [47] may be VEGF-dependent. Systemic anti-VEGF treatment for cancer has been reported to induce remission of cutaneous manifestations of psoriasis coincident with the cancer [47]. Conventional treatments for both of these conditions are inconsistently effective and systemic bevacizumab (anti-VEGF) treatment is too expensive to consider in most patients with skin disease.

Topical application of tetrac in a vehicle that permits penetration of the agent to involved blood vessels in the dermis and limits systemic absorption of the hormone analogue has been proposed in management of rosacea and awaits clinical trial.

Retinopathy

Tetrac and nanotetrac have been tested for efficacy in the newborn mouse oxygen-induced retinopathy (OIR) model [34]. It was an effective intravitreal or intraperitoneal preventive intervention. Similar results in this model have been obtained by S.A. Mousa et al. (unpublished). Yoshida and co-workers also found that the effects of VEGF and erythropoietin (EPO) on retinal endothelial cells *in vitro* were blocked by tetrac and nanotetrac [34]. As noted earlier, EPO is another factor supporting angiogenesis whose activity is minimized by tetrac.

VEGF antibody administered by the intravitreal route has been examined for its effectiveness in the clinical setting of diabetic retinopathy [44, 48]. The substantial experience is largely favorable. The increased intravitreous and circulating levels of VEGF seen in proliferative retinopathy are both decreased by bevacizumab [49]. The agent may, however, increase the risk of retinal fibrosis [50].

Inflammation

Analysis of the gene signature of tetrac treatment in relatively chemoresistant human breast cancer cells [15] has revealed an important set of actions on inflammation-related genes [51]. For example, five of six differentially regulated interleukin genes—including *IL-6* and *IL-1α*—are down-regulated by the compound and a suppressor of cytokine signaling (*SOCS4*) gene is up-regulated. Expression of interferon response pathway genes and chemokine genes is decreased. Such effects may be relevant to inflammation-associated cancers, as well as to other inflammatory states. The selectivity of the tetrac/nanotetrac effect on gene expression is shown in the case of *IL-11*, whose gene product is a proliferative factor for hematopoietic stem cells and whose expression is enhanced, rather than decreased, by nanotetrac and tetrac.

Additional Actions of Tetrac and Nanotetrac

Independently of their actions of angiogenesis, tetrac and nanotetrac inhibit proliferation of a variety of cancer cell lines *in vitro*, as noted above, and chemosensitize [19] and radiosensitize [20, 21] tumor cells. The mechanism of chemosensitization by tetrac is incompletely understood, but may involve acidification of the cancer cell by inhibition of the Na^+/H^+ exchanger [52]. Consequent increase in extracellular pH (pHe) or decrease in intracellular pH may be associated with decreased activity of the cancer cell P-glycoprotein (P-gp) or other multidrug resistance (MDR) pumps that export chemotherapeutic agents [53]. Tetrac may also interfere with stimulation of the cancer cell kinases relevant to MDR pump activation [54].

As noted earlier, the mechanism of radiosensitization of cancer cells involves inhibition of repair of double-strand DNA breaks induced by radiation [20]. Under normal conditions, this repair process is highly efficient in tumor cells.

Conclusions

Acting at the thyroid hormone-tetrac receptor on the plasma membrane, tetrac and nanotetrac have potentially important anti-angiogenic effects. The receptor is located on integrin $\alpha v\beta 3$, a highly plastic protein that is capable of transducing interactions of its extracellular domain with extracellular matrix proteins and small molecules, like thyroid hormone, into important intracellular events. Tetrac and its nanoparticulate formulation disrupt the communication between the integrin and nearby receptors for VEGF, bFGF, and other polypeptide factors important to neovascularization. That is, these thyroid hormone derivatives affect the activity of a number of pro-angiogenic proteins, in addition to blocking the angiogenic activity of agonist thyroid hormones, T_4 and T_3. Tetrac and nanotetrac also stimulate expression of the endogenous angiogenic suppressor gene, *thrombospondin 1*. The actions of tetrac and nanotetrac are generally coherent, that is, the effects that they have on expression of multiple genes, on crosstalk between integrin $\alpha v\beta 3$ and receptors such as VEGFR and bFGFR, and on vascular growth factor release by tumor cells fit an anti-angiogenic pattern. That the agents distinguish between endothelial cell Ang-1 and Ang-2 also supports coherence of the anti-angiogenic pharmacology of tetrac and nanotetrac.

These agents have potential applications as inhibitors of tumor-related angiogenesis and of angiogenesis that is not associated with malignancy, but contributes to clinically significant skin disorders, to vascular proliferation of the retina, and neovascularization associated with inflammatory states.

References

1. Moreno M, de Lange P, Lombardi A, Silvestri E, Lanni A, Goglia F (2008) Metabolic effects of thyroid hormone derivatives. Thyroid 18(2):239–253
2. Burger AG, Engler D, Sakaloff C, Staeheli V (1979) The effects of tetraiodothyroacetic and triiodothyroacetic acids on thyroid function in euthyroid and hyperthyroid subjects. Acta Endocrinol (Copenhagen) 92(3):455–467
3. Davis PJ, Handwerger BS, Gregerman RI (1972) Thyroid hormone binding by human serum prealbumin (TBPA). Electrophoretic studies of triiodothyronine-TBPA interaction. J Clin Invest 51(3):515–521

4. Davis FB, Cody V, Davis PJ, Borzynski LJ, Blas SD (1983) Stimulation by thyroid hormone analogues of red blood cell Ca^{2+}-ATPase activity in vitro. Correlation between hormone structure and biologic activity in a human cell system. J Biol Chem 258(20):12373–12377
5. Davis PJ, Davis FB, Lawrence WD, Blas SD (1989) Thyroid hormone regulation of membrane Ca(2+)-ATPase activity. Endocr Res 15(4):651–682
6. Warnick PR, Davis PJ, Davis FB, Cody V, Galindo J Jr, Blas SD (1993) Rabbit skeletal muscle sarcoplasmic reticulum Ca(2+)-ATPase activity: stimulation in vitro by thyroid hormone analogues and bipyridines. Biochim Biophys Acta 1153(2):184–190
7. Nieman LK, Davis FB, Davis PJ, Cunningham EE, Gutman S, Blas SD, Schoenl M (1983) Effect of end-stage renal disease on responsiveness to calmodulin and thyroid hormone of calcium-ATPase in human red blood cells. Kidney Int Suppl 16:S167–S170
8. Cheng SY, Leonard JL, Davis PJ (2010) Molecular aspects of thyroid hormone actions. Endocr Rev 31(2):139–170
9. Bergh JJ, Lin HY, Lansing L, Mohamed SN, Davis FB, Mousa S, Davis PJ (2005) Integrin alphav-beta3 contains a cell surface receptor for thyroid hormone that is linked to activation of mitogen-activated protein kinase and induction of angiogenesis. Endocrinology 146(7):2864–2871
10. Davis PJ, Davis FB, Mousa SA, Luidens MK, Lin HY (2011) Membrane receptor for thyroid hormone: physiologic and pharmacologic implications. Annu Rev Pharmacol Toxicol 51:99–115
11. Lin HY, Cody V, Davis FB, Hercbergs AA, Luidens MK, Mousa SA, Davis PJ (2011) Identification and functions of the plasma membrane receptor for thyroid hormone analogues. Discov Med 11(59):337–347
12. Gaertner FC, Schwaiger M, Beer AJ (2010) Molecular imaging of αvβ3 expression in cancer patients. Q J Nucl Med Mol Imaging 54(3):309–326
13. Freindorf M, Furlani TR, Kong J, Cody V, Davis FB, Davis PJ (2012) Combined QM/MM study of thyroid and steroid hormone analogue interactions with αvβ3 integrin. J Biomed Biotechnol Article ID 959057, doi:10.1155/2012/959057
14. Davis PJ, Davis FB, Lin HY, Mousa SA, Zhou M, Luidens MK (2009) Translational implications of nongenomic actions of thyroid hormone initiated at its integrin receptor. Am J Physiol Endocrinol Metab 297(6):E1238–E1246
15. Glinskii AB, Glinsky GV, Lin HY, Tang HY, Sun M, Davis FB, Luidens MK, Mousa SA, Hercbergs AH, Davis PJ (2009) Modification of survival pathway gene expression in human breast cancer cells by tetraiodothyroacetic acid (tetrac). Cell Cycle 8(21):3554–3562
16. Somananth PR, Malinin NL, Byzova TV (2009) Cooperation between integrin alphavbeta3 and VEGFR2 in angiogenesis. Angiogenesis 12(2):177–185
17. Mousa SA, Bergh JJ, Dier E, Rebbaa A, O'Connor LJ, Yalcin M, Aljada A, Dyskin E, Davis FB, Lin HY, Davis PJ (2008) Tetraiodothyroacetic acid, a small molecule integrin ligand, blocks angiogenesis induced by vascular endothelial growth factor and basic fibroblast growth factor. Angiogenesis 11(2):183–190
18. Davis FB, Mousa SA, O'Connor L, Mohamed S, Lin HY, Cao HJ, Davis PJ (2004) Proangiogenic action of thyroid hormone is fibroblast growth factor-dependent and is initiated at the cell surface. Circ Res 94(11):1500–1506
19. Rebbaa A, Chu P, Davis FB, Davis PJ, Mousa SA (2008) Novel function of the thyroid hormone analog tetraiodothyroacetic acid: a cancer chemosensitizing and anti-cancer agent. Angiogenesis 11(5):269–276
20. Hercbergs A, Davis PJ, Davis FB, Ciesielski MJ, Leith JT (2009) Radiosensitization of GL261 glioma cells by tetraiodothyroacetic acid (tetrac). Cell Cycle 8(16):2586–2591
21. Hercbergs AH, Lin HY, Davis FB, Davis PJ, Leith JT (2011) Radiosensitization and production of DNA double-strand breaks in U87MG brain tumor cells induced by tetraiodothyroacetic acid (tetrac). Cell Cycle 10(2):352–357
22. Bridoux A, Cui H, Dyskin E, Schmitzer AR, Yalcin M, Mousa SA (2010) Semisynthesis and pharmacological activities of thyroxine analogs: development of new angiogenesis modulators. Bioorg Med Chem Lett 20(11):3394–3398
23. Bharali DJ, Yalcin M, Davis PJ, Mousa SA (2012) Tetraiodothyroacetic acid (Tetrac) conjugated PLGA nanoparticles: a nanomedicine approach to treat drug-resistant breast cancer. Nanomedicine (in press), 2013. doi:10.2217/nnm.12.200

24. Yalcin M, Dyskin E, Lansing L, Bharali DJ, Mousa SS, Bridoux A, Hercbergs AH, Lin HY, Davis FB, Glinsky GV, Glinskii AB, Ma J, Davis PJ, Mousa SA (2010) Tetraiodothyroacetic acid (tetrac) and nanoparticulate tetrac arrest growth of medullary carcinoma of the thyroid. J Clin Endocrinol Metab 95(4):1972–1980

25. Zolotarjova NI, Hollis GF, Wynn R (2001) Unusually stable and long-lived ligand-induced conformations of integrins. J Biol Chem 276(20):17063–17068

26. Medina-Gomez G, Calvo RM, Obregon MJ (2008) Thermogenic effect of triiodothyroacetic acid at low doses in rat adipose tissue without adverse effects in the thyroid axis. Am J Physiol Endocrinol Metab 294(4):E688–E697

27. Sherman SI, Ringel MD, Smith MJ, Kopelen HA, Zoghbi WA, Ladenson PW (1997) Augmented hepatic and skeletal thyromimetic effects of tiratricol in comparison with levothyroxine. J Clin Endocrinol Metab 82(7):2153–2158

28. Klootwijk W, Friesema EC, Visser TJ (2004) A nonselenoprotein from amphioxus deiodinates triac but not T3: is triac the primordial bioactive thyroid hormone? Endocrinology 152(8):3259–3267

29. Folkman J (1995) Angiogenesis in cancer, vascular, rheumatoid and other disease. Nat Med 1(1):27–31

30. Folkman J (2007) Angiogenesis: an organizing principle for drug discovery? Nat Rev Drug Discov 6(4):273–286

31. Jensen RL (1998) Growth factor-mediated angiogenesis in malignant progression of glial tumors: a review. Surg Neurol 49(2):189–195

32. Ribbati D (2010) Erythropoietin and tumor angiogenesis. Stem Cells Dev 19(1):1–4

33. Mousa SA, Davis FB, Mohamed S, Davis PJ, Feng X (2006) Pro-angiogenesis action of thyroid hormone and analogs in a three-dimensional in vitro microvascular endothelial sprouting model. Int J Angiol 25(4):407–413

34. Yoshida T, Gong J, Xu Z, Wei Y, Duh EJ (2012) Inhibition of pathological angiogenesis by the integrin αvβ3 antagonist tetraiodothyroacetic acid (tetrac). Exp Eye Res 94(1):41–48

35. Yalcin M, Bharali DJ, Dyskin E, Dier E, Lansing L, Mousa SS, Davis FB, Davis PJ, Mousa SA (2010) Tetraiodothyroacetic acid and tetraiodothyroacetic acid nanoparticle effectively inhibit the growth of human follicular thyroid cell carcinoma. Thyroid 20(3):281–286

36. Yalcin M, Bharali DJ, Lansing L, Dyskin E, Mousa SS, Hercbergs A, Davis FB, Davis PJ, Mousa SA (2009) Tetraiodothyroacetic acid (tetrac) and tetrac nanoparticles inhibit growth of human renal cell carcinoma xenografts. Anticancer Res 29(10):3825–3831

37. Mousa SA, Yalcin M, Bharali DJ, Meng R, Tang HY, Lin HY, Davis FB, Davis PJ (2012) Tetraiodothyroacetic acid and its nanoformulation inhibit thyroid hormone stimulation of non-small cell lung cancer cells in vitro and its growth in xenografts. Lung Cancer 76(1):39–45

38. Borges E, Jan Y, Ruoslahti E (2000) Platelet-derived growth factor receptor beta and vascular endothelial growth factor receptor 2 bind to the beta3 integrin through its extracellular domain. J Biol Chem 275(51):39867–39873

39. Stuttfeld E, Ballmer-Hofer K (2009) Critical review. Structure and function of VEGF receptors. Life 61(9):915–922

40. Thomas M, Augustin HG (2009) The role of the angiopoietins in vascular morphogenesis. Angiogenesis 12(2):125–137

41. Nagy JA, Dvorak AM, Dvorak HF (2007) VEGF-A and the induction of pathological angiogenesis. Annu Rev Pathol 2:251–275

42. Shojaei F (2012) Antiangiogenesis therapy in cancer: current challenges and future perspectives. Cancer Lett 320(2):130–137

43. Ranpura V, Hapani S, Wu S (2011) Treatment-related mortality with bevacizumab in cancer patients: a meta-analysis. JAMA 305(5):487–494

44. Salam S, Mathew R, Sivaprasad S (2011) Treatment of proliferative diabetic retinopathy with anti-VEGF agents. Acta Ophthalmol 89(5):405–411

45. The IVAN Study Investigators Writing Committee, Chakravarthy U, Harding SP, Rogers CA, Downes SM, Lotery AJ, Wordsworth S, Reeves BC (2012) Ranibizumab versus bevacizumab to treat neovascular age-related macular degeneration: one-year findings from the IVAN Randomized Trial. Ophthalmology 119:1399–1411

46. Smith JR, Lanier VB, Braziel RM, Falkenhagen KM, White C, Rosenbaum JT (2007) Expression of vascular endothelial growth factor and its receptors in rosacea. Br J Ophthalmol 91(2):226–229
47. Canavese M, Altruda F, Ruzicka T, Schauber J (2010) Vascular endothelial growth factor (VEGF) in the pathogenesis of psoriasis—a possible target for novel therapies. J Dermatol Sci 58(3):171–176
48. Arevalo JF, Sanchez JG, Lasave AF, Wu L, Maia M, Bonafonte S, Brito M, Alezzandrini AA, Restrepo N, Berrocal MH, Saravia M, Farah ME, Fromow-Guerra J, Morales-Canton V (2010) Intravitreal bevacizumab (Avastin®) for diabetic retinopathy at 24-months: The 2008 Juan Verdaguer-Planas Lecture. Curr Diabetes Rev 6(5):313–322
49. Ma Y, Zhang Y, Zhao T, Jiang YR (2012) Vascular endothelial growth factor in plasma and vitreous fluid of patients with proliferative diabetic retinopathy patients after intravitreal injection of bevacizumab. Am J Ophthalmol 153(2):307–313
50. Van Geest RJ, Lesnik-Oberstein SY, Tan HS, Mura M, Goldschmeding R, Van Noorden CJ, Klaassen I, Schlingermann RO (2012) A shift in the balance of vascular endothelial growth factor and connective tissue growth factor by bevacizumab causes the angiofibrotic switch in proliferative diabetic retinopathy. Br J Ophthalmol 96(4):587–590
51. Lin HY, Glinsky GV, Glinskii AB, Davis FB, Mousa SA, Luidens MK, Hercbergs A, Davis PJ (2012) Tetraiodothyroacetic acid (tetrac) acts at a plasma membrane receptor to modulate expression of inflammation-related genes in tumor cells.In: 94th Annual meeting of The Endocrine Society, Houston, TX, 23–26 June, abstract 852243
52. D'Arezzo S, Incerpi S, Davis FB, Acconcia F, Marino M, Farias RN, Davis PJ (2004) Rapid nongenomic effects of 3, 5, 3′-triiodo-L-thyronine on the intracellular pH of L-6 myoblasts are mediated by intracellular calcium mobilization and kinase pathways. Endocrinology 145(12):5694–5703
53. Wojtkowiak JW, Verduzco D, Schramm KJ, Gillies RJ (2011) Drug resistance and cellular adaptation to tumor acidic pH microenvironment. Mol Pharm 8(6):2032–2038
54. Hait WN, Aftab DT (1992) Rational design and pre-clinical pharmacology of drugs for reversing multidrug resistance. Biochem Pharmacol 43(1):103–107

Chapter 11
Integrin Antagonists and Angiogenesis

Shaker A. Mousa and Paul J. Davis

Abstract Integrins and associated extracellular matrix protein ligands participate in angiogenesis, thrombosis, apoptosis, cell migration and proliferation. Disorders of such processes lead to acute and chronic disease states such as ocular diseases, cancer metastasis, unstable angina, myocardial infarction, stroke, osteoporosis, a wide range of inflammatory diseases, vascular remodeling and neurodegenerative disorders. Progress has been substantial in the development of antagonists for $\alpha v \beta 3$, $\alpha v \beta 5$, and $\alpha v \beta 1$ integrins to modulate angiogenesis and blood vessel-related disorders. Several reports illustrate existence of crosstalk between integrins and various hormonal systems. The expression of αv integrin on distinct cell types contributes to cancer growth, and αv integrin antagonists have the potential to disrupt multiple aspects of cancer and blood vessel disease progression. The rationale for the development of various therapeutic and diagnostic candidate anti-integrin agents is reviewed here, as are nanoparticle delivery systems directed at specific sites on integrins.

S.A. Mousa (✉)
The Pharmaceutical Research Institute at Albany College of Pharmacy
and Health Sciences, Rensselaer, NY, USA
e-mail: shaker.mousa@acphs.edu

P.J. Davis
The Pharmaceutical Research Institute at Albany College of Pharmacy
and Health Sciences, Rensselaer, NY, USA

Department of Medicine, Albany Medical College, Albany, NY, USA
e-mail: pdavis.ordwayst@gmail.com

S.A. Mousa and P.J. Davis (eds.), *Angiogenesis Modulations in Health and Disease:*   119
Practical Applications of Pro- and Anti-angiogenesis Targets,
DOI 10.1007/978-94-007-6467-5_11, © Springer Science+Business Media Dordrecht 2013

Integrins

Integrins are a widely expressed family of cell adhesion receptor proteins by which cells attach to extracellular matrices, or to adjacent cells. Integrins are heterodimeric proteins composed of alpha (α) and beta (β) monomers (Table 11.1). Most cells express several integrins. The interaction of plasma membrane integrins with the cytoskeleton inside the cell and through their extracellular domains with extracellular matrix appears to require the presence of both subunits. The binding of integrins to their ligands is a cation-dependent event. Integrins appear to recognize specific amino acid sequences in their ligands. The most well-studied is the Arg-Gly-Asp (RGD) sequence found within a number of matrix proteins, including fibrinogen, vitronectin, fibronectin, thrombospondin, osteopontin, von Willebrand factor (vWF) and others [1–3]. However, integrins may bind to ligands via a non-RGD binding domain, such as the $\alpha4\beta1$ integrin receptors that bind and recognize the LDV sequence within the CS-1 (connecting segment) region of fibronectin. There are at least 8 known α subunits and 14 β subunits [4–7]. Although the association of the different α and β subunits could in theory result in more than 100 integrins, the actual diversity is much more restricted.

Integrins and Signaling

Integrin adhesion receptors contain an extracellular face that engages adhesive ligands and a cytoplasmic face that engages intracellular proteins. The interactions between the cell adhesion molecules and extracellular matrix proteins are critical for cell adhesion and for anchorage-dependent signaling reactions in normal and

Table 11.1 αv Integrins: ligands and cellular and tissue distribution

αv Integrins	Ligands	Cellular and tissue distribution
αvβ1	Ln, Fn, Opn	Smooth muscle cells
		Fibroblasts, osteoclasts
		Tumor cells
αvβ3	Opn, Fg, Vn, Tn, TSP, Fn, PECM, MMP-2	Endothelial cells
		Smooth muscle cells
		Osteoclasts, platelets, fibroblasts
		Tumor cells, epithelial cells, leukocytes
αvβ5	Opn, Fg, Vn, Fn, TSP	Endothelial cells
		Smooth muscle cells
		Osteoclasts, platelets, fibroblasts
		Tumor cells, epithelial cells, leukocytes
αvβ6	Fn, Fg, Vn, Tn	Epithelial cells, carcinoma cells
αvβ8	Vn	Melanoma, kidney, brain, ovary, uterus, placenta

Abbreviations: *Fg* fibrinogen, *Fn* fibronectin, *Tn* tenascin, *TSP* thrombospondin, *Vn* vitronectin, *Opn* osteopontin, *Ln* laminin pathological states [8–13]. For example, platelet activation induces a conformational change in integrin αIIb/β3, thereby converting it into a high affinity fibrinogen receptor. Fibrinogen binding then triggers an activation cascade of protein tyrosine kinases and phosphatases and recruitment of numerous other signaling molecules into F-actin-rich cytoskeleton assemblies in proximity to the cytoplasmic tails of αIIb and β3 [14]. These dynamics appear to influence platelet function by co-coordinating signals emanating from integrins and G-protein linked receptors [14].

Studies of integrin mutations confirm that the cytoplasmic tails of αIIb/β3 are involved in integrin signaling, apparently through direct interactions with cytoskeleton and signaling molecules [15]. Blockade of fibrinogen binding to the extracellular face of αIIb/β3 has been shown to be an effective way to prevent platelet-rich arterial thrombi after coronary angioplasty in myocardial infarction and unstable angina pectoris patients [16–18]. Once proteins that interact with the cytoplasmic tails of αIIb/β3 are fully identified, it may also be possible to develop selective inhibitors of integrin adhesion or signaling whose locus of action is inside the cell [19]. This type of intracellular approach in modulating integrin function will perhaps be more difficult than achieving direct blockade of the integrins' extracellular binding because of the lack of cellular specificity for the integrin cytoskeleton coupled intracellular site(s).

Surface Membrane Integrin as a Potential Drug Discovery Target

A number of physiological processes, including cell activation, migration, proliferation, differentiation and many other processes, require direct contact between cells and extracellular matrix. Cell-cell and cell-matrix interactions are mediated through several different families of cell adhesion molecules (CAM), including the selectins, the integrins, the cadherins and the immunoglobulins. The commercial and therapeutic potential of cell adhesion molecules is on the rise [20, 21].

Newly discovered CAMs, along with the discovery of new roles for integrins, selectins and immunoglobulins in certain disease states, provide great opportunities to develop novel therapeutic and diagnostic modalities.

β1 Integrins

α2β1 Integrin

The α2β1 integrin-mediated binding to collagen I fibrils, but not that of α1β1, is involved in the regulation of angiogenesis-mediated processes as demonstrated with a small molecular weight α2β1 inhibitor [22]. Earlier studies showed that another α2β1 integrin blocker inhibited tumor angiogenesis and tumor growth [23].

α3β1 Integrin

The integrin α3β1 plays important roles in development, angiogenesis, and the pathogenesis of cancer, suggesting potential therapeutic uses for antagonists of this receptor. Recently, an α3β1 integrin-binding site was mapped to residues 190–201 (FQGVLQNVRFVF) of the N-terminal domain of the secreted protein thrombospondin-1 (TSP-1) [24]. The NVR motif is a required element of full-length TSP-1 for specific molecular recognition by the α3β1 integrin and biological activity.

α5β1 Integrin

α5β1 Integrin in Angiogenesis

Expression of the extracellular matrix protein fibronectin in provisional vascular matrices precedes permanent collagen expression and provides signals to vascular cells and fibroblasts during blood clotting and wound-healing, atherosclerosis and hypertension [25]. Fibronectin expression is also up-regulated in blood vessels in granulation tissues during wound-healing [26]. In fact, one isoform of fibronectin, the ED-B splice variant, is preferentially expressed on blood vessels in fetal and tumor tissues, but not on normal quiescent adult blood vessels [27]. These observations suggest a possible role for this isoform of fibronectin in angiogenesis. Animals lacking fibronectin were shown to die early in development from various defects, including missing notochord and somites as well as an improperly formed vasculature [28]. The functional role for fibronectin in vasculogenesis or in angiogenesis has never been directly established. However, Klein et al. concluded that engagement of the α5β1 integrin activates an NF-κB-dependent program of gene expression that coordinately regulates angiogenesis and inflammation [29].

Evidence was recently provided that both fibronectin and its receptor integrin α5β1 directly regulate angiogenesis [30, 31] and that interaction of fibronectin and α5β1 is central to the contribution of the two molecules to angiogenesis. In addition, integrin α5β1 participates in the same pathways of angiogenesis as integrin αvβ3, and these pathways are distinct from those involving integrin αvβ5 [30, 31]. Thus, antagonists for α5β1 integrin might be useful tools for the inhibition of angiogenesis associated with human tumor growth and neovascular-related ocular and inflammatory diseases [30, 31].

Integrin α5β1 plays an important role in developmental angiogenesis, but its role in various types of pathologic neovascularization has not been completely defined. Up-regulation of α5β1 in choroidal neovascularization has been demonstrated [32].

Implantation of an osmotic pump delivering approximately 1.8 or 12 mg/kg/day of 3-(2-{1-alkyl-5-[(pyridin-2-ylamino)-methyl]-pyrrolidin-3-yloxy}-acetyl amino) -2-(alkyl amino)-propionic acid (JSM6427), a selective α5β1 antagonist, caused significant suppression of choroidal neovascularization; the area of neovascularization was reduced by 33–40 %. Data from this investigation suggest that α5β1 plays a role in the development and maintenance of choroidal neovascularization and provides a target for therapeutic intervention [32]. Table 11.1 provides examples of α5β1 ligands. Potent anti-α5β1 monoclonal antibody and small molecule antagonists [33, 34] have been designed (Table 11.2).

Table 11.2 Examples of α5β1 integrin antagonists

Structures	Activities	References
	IC_{50} (nM) = 0.02[a]	[35]
	IC_{50} (nM) = 0.02[b]	[36]
	IC_{50} (nM) = 0.04[b]	[36]

(continued)

Table 11.2 (continued)

Structures	Activities	References
	IC_{50} (nM) = 0.06[b]	[36]

[a]The binding IC50 on integrin was determined by an α5β1-vitronectin ELISA assay
[b]α5β1-integrin mediated cell adhesion to fibronectin in the presence of Model-B like ligands

β3 Integrins

Integrin αvβ3

Integrins form cytoplasmic complexes with Src family kinases, cytoskeletal proteins, growth factor receptors, MAPK, Ras, the nuclear factor of kappa light polypeptide gene enhancer in B cells (NF-κB), PIP3K, and protein kinase C [37]. Although the relationship between integrins and the extracellular matrix is complex and not completely understood, growing evidence points to a role of αvβ3 integrin receptors in growth regulation and anti-apoptosis of tumor cells. The role of integrins in angiogenesis is also supported by the fact that integrins are active in angiogenic endothelium and dormant on quiescent endothelial cells, and that blockade of αvβ3 receptors decreases angiogenesis, causes tumors to regress, and triggers endothelial apoptosis. The αv family of integrins associates with various β subunits and interacts with diverse extracellular matrix proteins that have different functional implications (Table 11.1).

Integrin αvβ3 and Matrix Proteins in Vascular Remodeling

Vascular remodeling processes play key roles in the pathological mechanisms of atherosclerosis and restenosis. In response to vascular injury, such as by percutaneous transluminal coronary angioplasty (PTCA), matrix proteins like osteopontin and vitronectin are locally and rapidly up-regulated [38]. Osteopontin stimulates smooth muscle cell migration *via* its action on integrin αvβ3, and thereby contributes to neointima formation and restenosis [38]. Thus, specific matrix proteins acting *via* selected integrins, and especially αvβ3, may be important targets for selective antagonists aimed at blocking the pathological processes of restenosis [38, 39]. Examples of αvβ3 antagonists [40–54] are given in Table 11.3.

Table 11.3 Examples of $\alpha v \beta 3$ integrin antagonists

Structures	Activities	References
	IC_{50} (nM) = 0.07[a]	[55]
	IC_{50} (nM) = 0.1[a]	[55]
	IC_{50} (nM) = 0.1[a]	[56]
	IC_{50} (nM) = 0.08[a]	[57, 58]
	IC_{50} (nM) = 0.00092 (K562)[d]	[59]
	IC_{50} (nM) = 0.03 (K562)[d]	[33]

(continued)

Table 11.3 (continued)

Structures	Activities	References
	IC_{50} (nM) = 0.07[a]	[60]
	IC_{50} (nM) = 0.08[a]	[61]
	IC_{50} (nM) = 0.1[b]	[62]
	IC_{50} (nM) = 0.1[b]	[63]
	IC_{50} (nM) = 0.9[a]	[60]
	IC_{50} (nM) = 0.05[c]	[64]

(continued)

Table 11.3   (continued)

Structures	Activities	References
	IC_{50} (nM) = 0.1[b]	[65]
	IC_{50} (nM) = 0.03 (NCI-H1975)[d]	[66]

[a] The binding IC_{50} on integrin was determined by a scintillation-proximity assay (SPAV)
[b] The binding IC_{50} on integrin was determined by a solid phase receptor binding assays (SPRA)
[c] The binding IC_{50} on integrin was determined by an αvβ3-vitronectin ELISA assay
[d] The IC_{50} values denote the concentration of compounds necessary to inhibit initial cell attachment to extracellular coated substrates to 50 % of control

A selective integrin αvβ3 inhibitor, cRGDfV, improves outcomes in the middle cerebral artery occlusion model by preserving the blood–brain barrier, which mechanistically may occur in a VEGF- and VEGF receptor-mediated manner [67]. A series of pyrazole and isoxazole analogues as antagonists of the αvβ3 receptor showed low to sub-nanomolar potency against αvβ3, as well as good selectivity against αIIbβ3. Several compounds showed good pharmacokinetic properties in rats, in addition to anti-angiogenic activity in a mouse corneal micropocket model [68].

Integrin αvβ3 Antagonists Promote Tumor Regression by Inducing Apoptosis of Angiogenic Blood Vessels

A single intravascular injection of a cyclic peptide or monoclonal antibody antagonist of integrin αvβ3 disrupts ongoing angiogenesis on the chick choriallantoic membrane [69]. This leads to the rapid regression of histologically distinct human tumors transplanted onto this membrane. In fact, αvβ3 antagonists also prevent the spontaneous pulmonary metastasis of human melanoma cells [35, 36, 70–74]. All human tumors examined in this model are αvβ3 negative, which suggests that these antagonists have no direct effect on the tumor cells. Induction of angiogenesis by a tumor or cytokine promotes entry of vascular cells into the cell cycle and expression of integrin $\alpha v\beta 3$. After angiogenesis is initiated, antagonists of this integrin induce apoptosis of angiogenic vascular cells, leaving pre-existing quiescent blood vessels unaffected [71]. These studies are supported by *in vitro* results. Specifically, cultured human endothelial cells are protected from apoptosis when they are allowed to attach to immobilized anti-$\alpha v\beta 3$ monoclonal antibody LM-609 [71]. The adhesion event appears to decrease expression of p53 and Bax while increasing that of Bcl-2. Ligation of $\alpha v\beta 3$ is required for the survival and maturation of newly forming blood vessels, an event essential for the proliferation and metastatic properties of human tumor [75–77]. Integrin $\alpha v\beta 3$ is preferentially expressed on blood vessels undergoing angiogenesis. Antibody or peptide antagonists of this integrin block angiogenesis in response to human tumors or purified cytokines in several preclinical models. These inhibitors of $\alpha v\beta 3$ promote selective apoptosis of newly sprouting vessels, preventing their maturation. These findings indicate that antibody or peptide antagonists of integrin $\alpha v\beta 3$ may have therapeutic value in the treatment of diseases associated with angiogenesis [78].

Use of $\alpha v\beta 3$ for Targeted Delivery

In order to selectively block NF-κB-dependent signal transduction in angiogenic endothelial cells, an $\alpha v\beta 3$ integrin-specific adenovirus encoding dominant negative IκB (dnIκB) as a therapeutic gene was constructed [79]. RGD-targeted adenovirus delivered the dnIκB via $\alpha v\beta 3$ to become functionally expressed, leading to complete abolition of TNF-α-induced up-regulation of E-selectin, ICAM-1, VCAM-1, IL-6, IL-8, VEGF-A and Tie-2 [79]. The approach of targeted delivery of dnIκB into endothelial cells might be employed for diseases such as rheumatoid arthritis and inflammatory bowel disease where activation of NF-κB activity should be locally restored to basal levels in the endothelium.

Integrin $\alpha v\beta 3$ has been implicated in multiple aspects of tumor progression and metastasis. Many tumors have high expression of $\alpha v\beta 3$ that correlates with tumor progression. Therefore, $\alpha v\beta 3$ receptor is an excellent target for drug design and delivery. A number of high affinity small-molecule $\alpha v\beta 3$ antagonists conjugated to chemotherapy for selective delivery to $\alpha v\beta 3$ positive metastatic cancer cells are reported [54, 80–82].

Combination of $\alpha v\beta 3$ Antagonists and Chemotherapy/Radiotherapy

Combination of anti-integrin $\alpha v\beta 3$ therapy and other therapeutic approaches (such as chemotherapy, radiotherapy and gene therapy) has also been applied to cancer treatment. Mounting evidence suggests that there is potentially a synergistic effect of combined therapeutic approaches. Integrin $\alpha v\beta 3$ is expressed at low levels on epithelial cells and mature endothelial cells, but it is overexpressed on the activated endothelial cells of tumor neovasculature and some tumor cells. The increased expression of integrin αvβ3 during tumor growth, invasion, and metastasis presents an interesting molecular target for both early detection and treatment of rapidly growing solid tumors. In the past decade, many radiolabeled linear and cyclic RGD peptide antagonists have been evaluated as integrin αvβ3 targeted radiotracers. Significant progress has been made on their use for imaging tumors of different origin by single photon emission computed tomography (SPECT) or positron emission tomography (PET) in several tumor-bearing animal models. [18F]Galacto-RGD is under clinical investigation as the first integrin αvβ3 targeted radiotracer for noninvasive visualization of the activated integrin αvβ3 in cancer patients. Radiolabeled multimeric cyclic RGD peptides (dimers and tetramers) are useful as radiotracers to image tumor integrin αvβ3 expression by SPECT and PET, and some fundamental aspects for the development of integrin αvβ3 targeted radiotracers [83]. These include the choice of radionuclide and bifunctional chelators, selection of targeting biomolecules, and factors influencing integrin αvβ3 binding affinity and tumor uptake, as well as different approaches for modification of radiotracer pharmacokinetics.

Integrins αvβ3 and αvβ5 are important in tumor growth and angiogenesis and have been recently explored as targets for cancer therapy (Table 11.4). Radiotherapy also inhibits tumor growth and affects vasculature [87]. Authors explored the combination of integrin antagonist cilengitide (EMD 121974) and ionizing radiation. Radiation induces expression of αvβ3 integrin in endothelial and non-small-cell lung cancer models, and integrin antagonist cilengitide is a radiosensitizer in proportion to the levels of target integrin expression [87].

Anti-αv integrin monoclonal antibody (17E6) and the small molecule αvβ3/αvβ 5 integrin inhibitor (EMD121974) suppressed invasion and metastasis induced by CYR61 and attenuated metastasis of tumors growing within a pre-irradiated field [88]. αv integrin inhibition can thus be identified as a potential therapeutic approach for preventing metastasis in patients at risk for post-radiation recurrences.

Antagonists of Integrins αvβ3 and αvβ5 Inhibit Angiogenesis

The enhanced expression of αvβ3 during angiogenesis suggests that it plays a critical role in the angiogenic process. Recent experimental evidence supports this notion. Specifically, antagonists of integrin αvβ3 potently inhibit angiogenesis in a number of animal models. When angiogenesis is induced in the chick chorioallantoic membrane with purified cytokines, αvβ3 expression is stimulated by 4-fold within 72 h [70]. Topical application of LM-609, a monoclonal antibody antagonist of αvβ3, inhibits angiogenesis, but other anti-integrin antibodies do not inhibit angiogenesis [70]. Application of LM-609 or cyclic RGD peptide antagonists–but not of the other anti-integrin antibodies or control peptides–to tumors grown on the surface of CAMs reduces the growth of blood vessels into the tumor tissue. LM-609 has no effect on pre-existing vessels [70]. These findings suggest that αvβ3 plays a role in a late event of blood vessel formation that is common to embryonic neovascularization and angiogenesis. Antagonists of integrin αvβ3 inhibit the

Table 11.4 Examples of dual αvβ3 and αvβ5 integrin antagonists

Structures	Activities	References
	$IC_{50} \; \alpha_v\beta_3 \; (nM) = 0.3$ $IC_{50} \; \alpha_v\beta_5 \; (nM) = 0.2$	[84]
	$IC_{50} \; \alpha_v\beta_3 \; (nM) = 0.2$ $IC_{50} \; \alpha_v\beta_5 \; (nM) = 0.1$	[85]
	$IC_{50} \; \alpha_v\beta_3 \; (nM) = 0.2$ $IC_{50} \; \alpha_v\beta_5 \; (nM) = 0.2$	[86]
	$IC_{50} \; \alpha_v\beta_3 \; (nM) = 0.3$ $IC_{50} \; \alpha_v\beta_5 \; (nM) = 0.2$	[115]
	$IC_{50} \; \alpha_v\beta_3 \; (nM) = 0.2$ $IC_{50} \; \alpha_v\beta_5 \; (nM) = 0.2$	[115]

The binding IC_{50} on integrins was determined by an αvβ3 or αvβ5-vitronectin ELISA assay

growth of new blood vessels into tumors cultured on the chick chorioallantoic membrane without affecting adjacent blood vessels; they also induce tumor regression. Up to 5-fold differences in tumor sizes are observed between treated and control tumors. A single intravascular injection of LM-609 halts the growth of tumors and induces the regression of tumors as determined by tumor weight [35, 36, 70–72]. Similarly, an injection of a cyclic RGD peptide antagonist of αvβ3, but not of an inactive control peptide, induces tumor regression. Histological examination of the anti-αvβ3 and control treated tumor reveals that few, if any, viable tumor cells remain in the anti-αvβ3 treated tumors. In fact, these treated tumors contained no viable blood vessels [72].

Antagonists of integrin αvβ3 also inhibit tumor growth in human skin. In exciting studies of the effect of these antagonists on human angiogenesis, Brooks and colleagues implanted human breast carcinoma cells in human skin grafted on SCID mice [71]. Tumor growth was either completely suppressed or was significantly inhibited by LM-609 antibody directed at αvβ3 when compared to mice treated with an anti-body control. Angiogenesis was significantly inhibited (by at least 75 %) in the LM-609 treated animals. Antagonists of integrin αvβ3 also inhibit angiogenesis in various ocular models. Integrin αvβ3 peptide antagonists also inhibit murine retinal neovascularization in an oxygen-induced model of ischaemic retinopathy [78].

SB-267268, a non-peptidic antagonist of the αvβ3 and αvβ5 integrins, attenuates angiogenesis in a murine model of retinopathy of prematurity (ROP) and alters the expression of VEGF and its second receptor [55]. Non-peptide inhibitors of αvβ3 and αvβ5 integrins are effective in ROP and may be a suitable anti-angiogenic therapy for other ischemic retinal pathologies.

The expression of these receptors in cytokine-stimulated blood vessels also suggests they may play roles in vascular proliferation and migration events associated with restenosis after angioplasty [38, 39].

Integrin αvβ3 Antagonists *Versus* Anti-αvβ3 and αvβ5

Since the recognition of at least two αv integrin pathways for cytokine-mediated angiogenesis [70, 72], mixed αvβ3 and αvβ5 antagonists might be proven to be more effective in certain indications as compared to a specific anti-αvβ3. These dual antagonists are illustrated in Table 11.4 [33, 56–59]. A small library of cyclic RGD penta-peptide mimics, including benzyl-substituted azabicycloalkane amino acids, was synthesized. Arosio et al. have reported one compound with affinity for both the αvβ3 and the αvβ5 integrins [56].

Role of αvβ3 in Osteoporosis

Several studies have shown that αvβ3 integrin is involved in the bone remodeling process [60, 89]. Osteoclasts express αvβ3 integrin generously, and the integrin binds to a variety of extracellular matrix proteins including vitronectin, osteopontin, and bone sialoprotein. Arg-Gly-Asp (RGD)-containing peptides, RGD-mimetic, and blocking antibodies to αvβ3 integrin were shown to inhibit bone resorption *in vitro* and *in vivo*, suggesting that this integrin may play an important role in regu-lating osteoclast function. Coleman et al. identified two potent and selective non-peptide antagonists of the αvβ3 receptor [40]. On the basis of the efficacy shown in an *in vivo* model of bone turnover following once-daily oral administration, these two compounds were selected for clinical development for the treatment of osteoporosis. A number of RGD-containing proteins, including osteopontin, bone sialoprotein, vitronectin and fibrinogen are known to bind to αvβ3 and regulate bone remodeling. On the other hand, antibodies to αvβ3, RGD-peptides and small molecule αvβ3 antagonists have been shown to be efficacious in models of bone resorption, providing strong evidence that inhibitors of this integrin would be useful agents for the treatment of osteoporosis. RGD analogues have been shown to inhibit the attachment of osteoclasts to bone matrix and to reduce bone resorptive activity *in vitro*. The cell surface integrin, αvβ3, appears to play a role in this process. Peptidomimetic antagonists of αvβ3, based on the RGD recognition sequence, have been synthesized and evaluated in several assay systems. These compounds inhibited the binding of vitronectin to isolated αvβ3, inhibited αvβ3-dependent cell adhesion to vitronectin, and reduced the hyper calcaemic response to parathyroid hormone in parathyroidectomized rats. RGD analogues may represent a new approach to modulating osteoclast-mediated bone resorption and may prove to be useful in the treatment of osteoporosis [61, 90].

αvβ3 Ligands

Issues in the Development of αvβ3 Antagonists as Therapeutics

A number of lead αvβ3 antagonists are under preclinical investigation [62–66]. However, a number of important issues are slowing development of these antagonists. Pharmacokinetic issues include achieving orally active compounds with high oral bioavailability and half-life. Pharmacodynamic issues include measuring *ex vivo* efficacy in order to predict the optimal dose required for clinical benefit. Another problem is attaining an optimal efficacy/safety ratio (high therapeutic index) so that pathological angiogenesis is opposed with minimal impact on physiological angiogenesis processes. A possible solution to the latter problem may be targeted delivery to tumor vasculature, as suggested recently [91].

The binding of lead compounds and drugs to human serum albumin (HSA) is a ubiquitous problem in drug discovery because it modulates the availability of the leads and drugs to their intended target and limits biological efficacy. In spite of nanomolar binding affinity of identified lead compounds to human αvβ3 and αvβ5 integrins, high HSA binding (>97 %) emerged as a limiting feature for these leads. Structure-activity HSA binding data of organic acids reported in the literature has demonstrated that the incorporation of polar groups into a given molecule can dramatically decrease the affinity for HSA. Among the compounds synthesized, 3-[5- [2-(5,6,7,8- tetrahydro [1,8]naphthyridin-2-yl) ethoxy]indo 1-1-yl] -3-[5-(N,N-dimethylaminomethyl)-3-pyridyl]propionic acid was found to be the most promising derivative within this novel series. It has a sub-nanomolar affinity for both αvβ3 and αvβ5 and has exhibited low HSA protein binding [92].

Diagnostics

Imaging metastatic cancer using technetium-99m-labeled RGD-containing synthetic peptide has been described. Detection of tumor angiogenesis *in vivo* by αvβ3-targeted magnetic resonance imaging (MRI) has also been demonstrated [93–95].

The $\alpha v \beta 3$ integrin is expressed in sprouting endothelial cells in growing tumors, whereas it is absent in quiescent blood vessels. In addition, various tumor cell types express this integrin. Due to the selective expression of $\alpha v \beta 3$ integrin in tumors, radiolabeled RGD peptides and peptidomimetics are attractive candidates for tumor targeting. A peptidomimetic compound and the cyclic RGD peptide have been shown to have high affinity for $\alpha v \beta 3$ integrin, and these compounds have good experimental (xenografted) tumor-targeting characteristics [96].

A series of radiolabeled cyclic RGD peptide ligands for cell adhesion molecule integrin $\alpha v \beta 3$-targeted tumor angiogenesis targeting are being developed [83]. This effort continues by applying a positron emitter ^{64}Cu-labeled PEGylated dimeric RGD peptide radiotracer ^{64}Cu-DOTA-PEG-E[c(RGDyK)]2 for lung cancer imaging [97]. The PEGylated RGD peptide showed integrin $\alpha v \beta 3$ avidity, but the PEGylation reduced the receptor binding affinity of this ligand compared to the unmodified RGD dimer. The radiotracer revealed rapid blood clearance and a predominantly renal clearance route. The minimum nonspecific activity accumulation in normal lung tissue and heart rendered high-quality orthotopic lung cancer tumor images, enabling clear demarcation of both the primary tumor at the upper lung lobe as well as metastases in the mediastinum, contralateral lung, and diaphragm. As a comparison, fluoro-deoxyglucose (FDG) scans on the same mice were only able to identify the primary tumor, with the metastatic lesions masked by intense cardiac uptake and high lung background. ^{64}Cu-DOTA-PEG-E[c(RGDyK)]2 is an excellent position emission tomography (PET) tracer for integrin-positive tumor imaging [97]. Further studies to improve the receptor binding affinity of the tracer and thereby to increase the magnitude of tumor uptake without comprising the favorable *in vivo* kinetics are currently in progress.

Integrin–Hormone Crosstalk

Integrin Receptor-mediated Actions of Thyroid Hormone

Evidence that thyroid hormone can act primarily outside the cell nucleus has come from studies of mitochondrial responses to T_3 or T_2 [98, 99], from rapid onset effects of the hormone at the cell membrane and from actions on cytoplasmic proteins [100–104]. The recent description of a plasma membrane receptor for thyroid hormone on integrin $\alpha v \beta 3$ [105–107] has provided some insight into effects of the hormone on membrane ion pumps, such as the Na^+/H^+ antiporter [102, 108], and has led to the description of interfaces between the membrane thyroid hormone receptor and nuclear events that underlie important cellular or tissue processes, such as angiogenesis and proliferation of certain tumor cells [109–112].

The possible clinical utility of cellular events that are mediated by the membrane receptor for thyroid hormone may reside in inhibition of such effect(s) in the contexts of neovascularization or tumor cell growth. Indeed, we have shown that blocking the membrane receptor for iodothyronines with tetraiodothyroacetic acid (tetrac), a thyroid hormone analogue that has no agonist activity at the receptor, can arrest growth of a variety of human cancer cells *in vitro* and in xenografts [112,113]. In recent studies, we examined the possibility of (tetrac) binding to the integrin receptor that may modulate vascular growth factor-induced angiogenesis in the absence of thyroid hormone. We concluded that tetrac acts via the integrin thyroid hormone receptor site to inhibit crosstalk between plasma-membrane vascular growth factor receptors and integrin $\alpha v \beta 3$ [114]. Tetrac is a useful probe to screen for participation of the integrin receptor in actions of thyroid hormone.

Integrin $\alpha v \beta 3$ binds thyroid hormone near the RGD recognition site of the protein, but these two sites are anatomically and functionally distinct [113]. The RGD site modulates protein-protein interactions linking the integrin to extracellular matrix proteins such as vitronectin, fibronectin and laminin [106]. The intact integrin is structurally very plastic [115]. Its conformational changes in response to ligand-binding may underlie its ability to transduce cell surface signals into discrete intracellular messages, as well as the ability to expose new surfaces for interactions. The integrin also generates crosstalk with other cell surface receptors. The thyroid hormone signal at the integrin is transduced into mitogen-activated protein kinase (MAPK) activity via phospholipase C and PKC [84]. MAPK (ERK1/2) activation is associated with increased Na^+/H^+ antiporter activity locally at the plasma membrane in response to thyroid hormone, and we speculate that hormone effects on other ion pumps at the cell surface relate to MAPK or PKC activation [108, 113]. Also initiated at the cell surface integrin receptor for thyroid hormone is the complex process of angiogenesis, monitored in either a standard chick blood vessel assay or with human endothelial cells in a sprouting assay [109]. This hormone-dependent process requires MAPK activation and elaboration of basic fibroblast growth factor (bFGF; FGF2) that are downstream mediators of thyroid hormone's effect on angiogenesis [109]. Tetrac blocks this pro-angiogenic action of T_4 and T_3, as does RGD peptide and small molecules that mimic RGD peptide. It is possible that desirable neovascularization can be promoted with local application of thyroid hormone analogues, e.g., in wound-healing, or that undesirable angiogenesis, such as that which supports tumor growth, can be antagonized with tetrac.

Thyroid hormone can also stimulate the proliferation *in vitro* of certain tumor cell lines [106, 112, 113]. Murine glioma cells [112] and human glioblastoma cells [116] have been shown to proliferate in response to physiological concentrations of T_4 by a mechanism initiated at the integrin receptor and that is MAPK-dependent. In what may be a clinical corollary, a prospective study of patients with far-advanced glioblastoma multiform in which mild hypothyroidism was induced by propylthiouracil showed an important survival benefit over euthyroid control patients [85]. We reported in 2004 that human breast cancer MCF-7 cells proliferated in response to T_4 by a mechanism that was inhibited by tetrac [117]. A recent retrospective clinical analysis by Cristofanilli et al. [86] showed that hypothyroid women who developed breast cancer did so later in life than matched euthyroid controls and had less aggressive, smaller lesions at the time of diagnosis than controls. Thus, the trophic action of thyroid hormone on *in vitro* models of both brain tumor and breast cancer appears to have clinical support.

Conclusions

The endothelial and smooth muscle cell αv integrin represents a potential strategy for targeted delivery in pathological angiogenesis and other vascular-mediated disorders. Additionally, the α5β1 integrin has been implicated in the modulation of angiogenesis, in a similar fashion to the αvβ3 integrin. Stimulated endothelial cells depend on αvβ3 function for survival during a critical period of the angiogenesis process, as inhibition of αvβ3/ligand interaction by antibody or peptide antagonists induces vascular cell apoptosis and inhibits angiogenesis. These observations open the door for further analysis of the regulation of cellular function and cell signaling by integrins, as well as for new therapeutic strategies to treat angiogenesis-associated disease.

These strategies have led to the development of antagonists to integrin αvβ3 and integrin αvβ5 that promote unscheduled programmed cell death of newly sprouting blood vessels. These antagonists cause regression of pre-established human tumors growing in laboratory animals and may thus lead to an effective therapeutic approach for solid tumors in humans. These antagonists include: (a) peptide inhibitors of individual integrins as well as peptides that inhibit integrins, (b) non-peptide, organic inhibitors, and (c) chimeric or humanized antibody inhibitors of integrin αvβ3. The first antagonist, a humanized form of the antibody LM-609 (Vitaxin), has already entered phase II clinical trials, and the first of the cyclic peptide antagonists is in initial clinical developments.

References

1. Hwang DS, Sim SB, Cha HJ (2007) Cell adhesion biomaterial based on mussel adhesive protein fused with RGD peptide. Biomaterials 28:4031–4046
2. Ruoslahti E (2003) The RGD, story: a personal account. Matrix Biol 22:459–465
3. Ruoslahti E, Pierschbacher M (1987) New perspectives in cell adhesion: RGD and integrins. Science 238:491–497
4. Ruoslahti E, Pierschbacher M (1986) ARG-GLY-ASP: a versatile cell recognition sequence. Cell 44:517–518
5. Hynes RO (1992) Integrins: versatility, modulation and signaling in cell adhesion. Cell 69:11–25
6. Cox D, Aoki T, Seki J, Motoyama Y, Yoshida K (1994) The pharmacology of the integrins. Med Res Rev 14(2):195–228
7. Albelda SM, Buck CA (1990) Integrins and other cell adhesion molecules. FASEB J 4:2868–2880
8. Cheresh D (1993) Integrins: structure, function and biological properties. Adv Mol Cell Biol 6:225–252
9. Guadagno TM, Ohtsubo M, Roberts JM et al (1993) A link between cyclin A expression and adhesion dependent cell cycle proliferation. Science 262:1572–1575
10. Juliano RL, Haskill S (1993) Signal transduction from the extracellular matrix. J Cell Biol 120:577–585
11. Kornberg LJ, Earb HS, Turner CE et al (1991) Signal transduction by integrins: increased protein tyrosine phosphorylation caused by clustering of β1 integrins. Proc Natl Acad Sci USA 88:8392–8395

12. Kornberg L, Earp HS, Parsons JT et al (1992) Cell adhesion or integrin clustering increased phorphorylation of a focal adhesion associated kinase. J Biol Chem 117:1101–1107
13. Guan JL, Salloway D (1992) Regulation of focal adhesion associated protein tyrosine kinase by both cellular adhesion and oncogenic transformation. Nature 358:690–692
14. Pelletier AJ, Bodary SX, Levinson AD (1992) Signal transduction by the platelet integrin αIIbβ3: induction of calcium oscillations required for protein-tyrosine phosphorylation and ligand-induced spreading of stably transfected cells. Mol Biol Cell 3:989–998
15. Vinogradova O, Velyvis A, Velyviene A, Hu B, Haas T, Plow E, Qin J (2002) A structural mechanism of integrin alpha(IIb)beta(3) "inside-out" activation as regulated by its cytoplasmic face. Cell 110(5):587–597
16. Topol EJ, Califf RM, Weisman HF et al (1994) Randomised trial of coronary intervention with antibody against platelet IIb/IIIa integrin for reduction of clinical restenosis: results at six months. Lancet 343:881–886
17. The EPIC Investigators (1994) Use of a monoclonal antibody directed against the platelet glycoprotein IIb/IIIa receptor in high-risk coronary angioplasty. The EPIC Investigation. N Engl J Med 330:956–961
18. Mousa SA (1999) Antiplatelet therapies: from aspirin to GPIIb/IIIa receptor antagonists and beyond. Drug Discov Today 4(12):552–561
19. Ma YQ, Qin J (2007) Plow EFPlatelet integrin alpha(IIb)beta(3): activation mechanisms. J Thromb Haemost 5(7):1345–1352
20. Staunton DE, Marlin SD, Stratowa C, Dustin ML, Springer TA (1988) Primary structure of ICAM-1 demonstrates interaction between members of the immunoglobulin and integrin supergene families. Cell 52:925–929
21. Romo GM, Dong J, Schade AJ, Gardiner EE, Kansas GS, Li CQ, McIntire LV, Berndt MC, Lopez JA (1999) The glycoprotein Ib-IX-V complex is a platelet counterreceptor for P-selectin. J Exp Med 190:803–814
22. San Antonio JD, Zoeller JJ, Habursky K, Turner K, Pimtong W, Burrows M, Choi S, Basra S, Bennett JS, DeGrado WF, Iozzo RV (2009) A key role for the integrin alpha2beta1 in experimental and developmental angiogenesis. Am J Pathol 175(3):1338–1347
23. Funahashi Y, Sugi NH, Semba T, Yamamoto Y, Hamaoka S, Tsukahara-Tamai N, Ozawa Y, Tsuruoka A, Nara K, Takahashi K, Okabe T, Kamata J, Owa T, Ueda N, Haneda T, Yonaga M, Yoshimatsu K, Wakabayashi T (2002) Sulfonamide derivative, E7820, is a unique angiogenesis inhibitor suppressing an expression of integrin alpha2 subunit on endothelium. Cancer Res 62(21):6116–6123
24. Furrer J, Luy B, Basrur V, Roberts DD, Barchi JJ Jr (2006) Conformational analysis of an alpha3beta1 integrin-binding peptide from thrombospondin-1: implications for antiangiogenic drug design. J Med Chem 49(21):6324–6333
25. Magnusson MK, Mosher DF (1998) Fibronectin: structure, assembly, and cardiovascular implications. Arterioscler Thromb Vasc Biol 18:1363–1370
26. Clark RA, Dellapelle P, Manseua E, Lanigan JM, Dvorak HF, Colvin RB (1982) Blood vessel fibronectin increases in conjunction with endothelial cell proliferation and capillary in growth during wound healing. J Invest Dermatol 79:269–276
27. Neri D, Carnimolla B, Nissim A et al (1997) Targeting by affinity-matured recombinant antibody fragments of an angiogenesis associated fibronectin isoform. Nat Biotechnol 15:1271–1275
28. George EL, Georges EN, Patel-King RS, Rayburn H, Hynes RO (1993) Defects in mesodermal migration and vascular development in fibronectin-deficient mice. Development 119:1079–1091
29. Klein S, de Fougerolles AR, Blaikie P, Khan L, Pepe A, Green CD, Koteliansky V, Giancotti FG (2002) α5β1 integrin activates an NF-κB-dependent program of gene expression important for angiogenesis and inflammation. Mol Cell Biol 22(16):5912–5922
30. Varner J, Mousa S (1998) Antagonists of vascular cell integrin α5β1 inhibit angiogenesis. Circulation 98(17 Suppl 1):1–795. 4166

31. Mousa S, Mohamed S, Sallhear J, Jadhav PK, Varner J (1999) Anti-angiogenesis efficacy of small molecule a5b1 integrin antagonist, SJ749. Blood 94(10 Suppl I):620a. 2755

32. Umeda N, Kachi S, Akiyama H, Zahn G, Vossmeyer D, Stragies R, Campochiaro PA (2006) Suppression and regression of choroidal neovascularization by systemic administration of an alpha5beta1 integrin antagonist. Mol Pharmacol 69(6):1820–1828

33. Raboisson P, Manthey CL, Chaikin M, Lattanze J, Crysler C, Leonard K, Pan W, Tomczuk BE, Marugán JJ (2006) Novel potent and selective alphavbeta3/alphavbeta5 integrin dual antagonists with reduced binding affinity for human serum albumin. Eur J Med Chem 41:847–861

34. Benfatti F, Cardillo G, Fabbroni S, Galzerano P, Gentilucci L, Juris R, Tolomelli A, Baiula M, Spartà A, Spampinato S (2007) Synthesis and biological evaluation of non-peptide alpha(v)beta(3)/alpha(5)beta(1) integrin dual antagonists containing 5,6-dihydropyridin-2-one scaffolds. Bioorg Med Chem 15:7380–7390

35. Albelda SM, Mette SA, Elder DE et al (1990) Integrin distribution in malignant melanoma: association of the b3 subunit with tumor progression. Cancer Res 50:6757–6764

36. Nip J, Brondt P (1995) The role of the integrin vitronectin receptor avb3 in melanoma metastasis. Cancer Metastasis Rev 14:241–252

37. Pechkovsky DV, Scaffidi AK, Hackett TL, Ballard J, Shaheen F, Thompson PJ, Thannickal VJ, Knight DA (2008) Transforming growth factor beta1 induces alphavbeta3 integrin expression in human lung fibroblasts via a beta3 integrin-, c-Src-, and p38 MAPK-dependent pathway. J Biol Chem 283(19):12898–12908

38. Srivatsa SS, Tsao P, Holmes DR, Schwartz RS, Mousa SA (1997) Selective $\alpha v\beta 3$ integrin blockade potently limits neointimal hyperplasia and lumen stenosis following deep coronary arterial stent injury. Cardiovasc Res 36:408–428

39. Zee R, Passeri J, Barry J, Cheresh D, Isner J (1996) A neutralizing antibody to the avb3 integrin reduces neointimal thickening in a balloon-injured iliac artery. Circulation 94(8):1505

40. Coleman PJ, Brashear KM, Askew BC, Hutchinson JH, McVean CA, Duong LT, Feuston BP, Fernandez-Metzler C, Gentile MA, Hartman GD, Kimmel DB, Leu C-T, Lipfert L, Merkle K, Pennypacker B, Prueksaritanont T, Rodan GA, Wesolowski GA, Rodan SB, Duggan ME (2004) Nonpeptide $\alpha v\beta 3$ antagonists. Part 11: discovery and preclinical evaluation of potent $\alpha v\beta 3$ antagonists for the prevention and treatment of osteoporosis. J Med Chem 47:4829–4837

41. Whitman DB, Askew BC, Duong LT, Fernandez-Metzler C, Halczenko W, Hartman GD, Hutchinson JH, Leu C-T, Prueksaritanont T, Rodan GA, Rodan SB, Duggan ME (2004) Nonpeptide $\alpha v\beta 3$ antagonists. Part 9: improved pharmacokinetic profile through the use of an aliphatic, des-amide backbone. Bioorg Med Chem Lett 14:4411–4415

42. Perkins JJ, Duong LT, Fernandez-Metzler C, Hartman GD, Kimmel DB, Leu C-T, Lynch JJ, Prueksaritanont T, Rodan GA, Rodan SB, Duggan ME, Meissner RS (2003) Non-peptide $\alpha v\beta 3$ antagonists: identification of potent, chain-shortened RGD mimetics that incorporate a central pyrrolidinone constraint. Bioorg Med Chem Lett 13:4285–4288

43. Breslin MJ, Duggan ME, Halczenko W, Hartman GD, Duong LT, Fernandez-Metzler C, Gentile MA, Kimmel DB, Leu C-T, Merkle K, Prueksaritanont T, Rodan GA, Rodan SB, Hutchinson JH (2004) Nonpeptide $\alpha v\beta 3$ antagonists. Part 10: in vitro and in vivo evaluation of a potent 7-methyl substituted tetrahydro-[1,8]naphthyridine derivative. Bioorg Med Chem Lett 14:4515–4518

44. Hutchinson JH, Halczenko W, Brashear KM, Breslin MJ, Coleman PJ, Duong LT, Fernandez-Metzler C, Gentile MA, Fisher JE, Hartman GD, Huff JR, Kimmel DB, Leu C-T, Meissner RS, Merkle K, Nagy R, Pennypacker B, Perkins JJ, Prueksaritanont T, Rodan GA, Varga SL, Wesolowski GA, Zartman AE, Rodan SB, Duggan ME (2003) Nonpeptide $\alpha v\beta 3$ antagonists. 8. In vitro and in vivo evaluation of a potent v3 antagonist for the prevention and treatment of osteoporosis. J Med Chem 46:4790–4798

45. Bubenik M, Meerovitch K, Bergeron F, Attardo G, Chan L (2003) Thiophene-based vitronectin receptor antagonists. Bioorg Med Chem Lett 13:503–506

46. Meerovitch K, Bergeron F, Leblond L, Grouix B, Poirier C, Bubenik M, Chan L, Gourdeau H, Bowlin T, Attardo G (2003) A novel RGD antagonist that targets both avß3 and a5ß1 induces apoptosis of angiogenic endothelial cells on type I collagen. Vascul Pharmacol 40:77–89
47. Cacciari B, Spalluto G (2005) Non peptidic αvβ3 antagonists: recent developments. Curr Med Chem 12:51–70
48. Ishikawa M, Kubota D, Yamamoto M, Kuroda C, Iguchi M, Koyanagi A, Murakami S, Ajito K (2006) Tricyclic pharmacophore-based molecules as novel integrin αvβ3 antagonists. Part 2: synthesis of potent αvβ3/αIIbβ3 dual antagonists. Bioorg Med Chem 14:2109–2130
49. Urbahns K, Härter M, Albers M, Schmidt D, Stelte-Ludwig B, Brüggemeier U, Vaupel A, Gerdes C (2002) Biphenyls as potent vitronectin receptor antagonists. Bioorg Med Chem Lett 12:205–208
50. Ishikawa M, Hiraiwa Y, Kubota D, Tsushima M, Watanabe T, Murakami S, Ouchi S, Ajito K (2006) Tricyclic pharmacophore-based molecules as novel integrin αvβ3 antagonists. Part III: synthesis of potent antagonists with αvβ3/αIIbβ3 dual activity and improved water solubility. Bioorg Med Chem 14:2131–2150
51. Penning TD, Russell MA, Chen BB, Chen HY, Desai BN, Docter SH, Edwards DJ, Gesicki GJ, Liang C-D, Malecha JW, Yu SS, Engleman VW, Freeman SK, Hanneke ML, Shannon KE, Westlin MM, Nickols GA (2004) Synthesis of cinnamic acids and related isosteres as potent and selective αvβ3 receptor antagonists. Bioorg Med Chem Lett 14:1471–1476
52. Iwama S, Kitano T, Fukuya F, Honda Y, Sato Y, Notake M, Morie T (2004) Discovery of a potent and selective αvβ3 integrin antagonist with strong inhibitory activity against neointima formation in rat balloon injury model. Bioorg Med Chem Lett 14:2567–2570
53. Lange UEW, Backfisch G, Delzer J, Geneste H, Graef C, Hornberger W, Kling A, Lauterbach A, Subkowski T, Zechel C (2002) Synthesis of highly potent and selective hetaryl ureas as integrin αvβ3-Receptor antagonists. Bioorg Med Chem Lett 12:1379–1382
54. Dayam R, Aiello F, Deng J, Wu Y, Garofalo A, Chen X, Neamati N (2006) Discovery of small molecule integrin αvβ3 antagonists as novel anticancer agents. J Med Chem 49:4526–4534
55. Wilkinson-Berka JL, Jones D, Taylor G, Jaworski K, Kelly DJ, Ludbrook SB, Willette RN, Kumar S, Gilbert RE (2006) SB-267268, a nonpeptidic antagonist of alpha(v)beta3 and alpha(v)beta5 integrins, reduces angiogenesis and VEGF expression in a mouse model of retinopathy of prematurity. Invest Ophthalmol Vis Sci 47(4):1600–1605
56. Arosio D, Belvisi L, Colombo L, Colombo M, Invernizzi D, Manzoni L, Potenza D, Serra M, Castorina M, Pisano C, Scolastico C (2008) A potent integrin antagonist from a small library of cyclic RGD pentapeptide mimics including benzyl-substituted azabicycloalkane amino acids. ChemMedChem 3(10):1589–1603
57. Perron-Sierra F, Saint Dizier D, Bertrand M, Genton A, Tucker GC, Casara P (2002) Substituted benzocyloheptenes as potent and selective αv integrin antagonists. Bioorg Med Chem Lett 12:3291–3296
58. Nadrah K, Dolenc MS (2005) Dual antagonists of integrins. Curr Med Chem 12:1449–1466
59. Raboisson P, DesJarlais RL, Reed R, Lattanze J, Chaikin M, Manthey CL, Tomczuk BE, Marugan JJ (2007) Identification of novel short chain 4-substituted indoles as potent antagonist using structure-based drug design. Eur J Med Chem 42:334–343
60. Hutchinson JH, Halczenko W, Brashear KM et al (2003) Nonpeptide alphavbeta3 antagonists. 8. In vitro and vivo evaluation of a potent alphavbeta3 antagonist for the prevention and treatment of osteoporosis. J Med Chem 46:4790–4798
61. Horton MA, Taylor ML, Arnett TR, Helfrich MH (1991) Arg-Gly-Asp (RGD) peptides and the antivitronectin receptor antibody 23C6 inhibit dentine resorption and cell spreading by osteoclasts. Exp Cell Res 195:368–375
62. Keenan RM, Miller WH, Kwon C et al (1997) Discovery of potent non-peptide vitronectin αvβ3 antagonists. J Med Chem 40:2289–2292
63. Corbett JW, Graciani NR, Mousa SA, Degrado WF (1997) Solid-phase synthesis of a selective αvβ3 integrin antagonist library. Bioorg Med Chem Lett 7:1371–1376
64. Knolle J, Breiphol G, Guba W et al (1997) Design and synthesis of potent and selective peptidomimetic vitronectin receptor antagonists. In: Tam JP, Kaumaya TP (eds) Peptide

chemistry structure and biology. Proceedings of the 15th American peptide symposium, Mayflower Scientific Ltd, Kingswinford, England, Abstract L102

65. Kerr JS, Wexler RS, Mousa SA et al (1999) Novel small molecule αv integrin antagonists: comparative anti-cancer efficacy with known angiogenesis inhibitors. Anticancer Res 19:959–968

66. Mousa SA, Lorelli W, Mohamed S, Batt DG, Jadhav PK, Reilly TM (1999) αvβ3 Integrin binding affinity and specificity of SM256 in various species. J Cardiovasc Pharmacol 33:641–646

67. Shimamura N, Matchett G, Yatsushige H, Ohkuma H, Zhang J (2006) Inhibition of integrin alphavbeta3 ameliorates focal cerebral ischemic damage in the rat middle cerebral artery occlusion model. Stroke 37(7):1902–1909

68. Penning TD, Khilevich A, Chen BB, Russell MA, Boys ML, Wang Y, Duffin T, Engleman VW, Finn MB et al (2006) Synthesis of pyrazoles and isoxazoles as potent alpha(v)beta3 receptor antagonists. Bioorg Med Chem Lett 16(12):3156–3161

69. Brooks PC, Clark RA, Cheresh DA (1994) Requirement of vascular integrin αvβ3 for angiogenesis. Science 264:569–571

70. Sanders LC, Felding-Habermann B, Mueller BM et al (1992) Role of αv integrins and vitronectin in human melanoma cell growth. Cold Spring Harb Symp Quant Biol 57:233–240

71. Brooks PC, Strömbland S, Klemke R et al (1995) Anti-integrin αvβ3 blocks human breast cancer growth and angiogenesis in human skin. J Clin Invest 96:1815–1822

72. Hieken TJ, Farolan M, Ronan SG et al (1996) β3 integrin expression in melanoma predicts subsequent metastasis. J Surg Res 63:169–173

73. Max R, Gerritsen R, Nooijen P et al (1997) Immunohistochemical analysis of integrin αvβ3 expression on tumor-associated vessels of human carcinomas. Int J Cancer 71:320–324

74. Clark RA, Tonnesen MG, Gailit J et al (1996) Transient functional expression of alphaVbeta 3 on vascular cells during wound repair. Am J Pathol 148:1407–1421

75. Eliceiri BP, Klemke R, Strömbland S, Cheresh DA (1999) Integrin αvβ3 requirement for sustained mitogenactivated protein kinase activity during angiogenesis. J Cell Biol 140:1255–1263

76. Friedlander M, Brooks PC, Shaffer RW et al (1995) Definition of two angiogenic pathways by distinct αv integrins. Science 270:1500–1502

77. Eliceiri BP, Cheresh DA (1999) The role of αv integrins during angiogenesis: insights into potential mechanisms of action and clinical development. J Clin Invest 103:1227–1230

78. Luna J, Tobe T, Mousa SA et al (1996) Antagonists of integrin αvβ3 inhibit retinal neovascularization in a murine model. Lab Invest 75:563–573

79. Ogawara K, Kułdo JM, Oosterhuis K, Kroesen BJ, Rots MG, Trautwein C, Kimura T, Haisma HJ, Molema G (2006) Functional inhibition of NF-kappaB signal transduction in alphavbeta3 integrin expressing endothelial cells by using RGD-PEG-modified adenovirus with a mutant IkappaB gene. Arthritis Res Ther 8(1):R32

80. Curnis F, Sacchi A, Gasparri A, Longhi R, Bachi A, Doglioni C, Bordignon C, Traversari C, Rizzardi GP, Corti A (2008) Isoaspartate-glycine-arginine: a new tumor vasculature-targeting motif. Cancer Res 68(17):7073–7082

81. Zhao H, Wang JC, Sun QS, Luo CL, Zhang Q (2009) RGD-based strategies for improving antitumor activity of paclitaxel-loaded liposomes in nude mice xenografted with human ovarian cancer. J Drug Target 17(1):10–18

82. Garanger E, Boturyn D, Dumy P (2007) Tumor targeting with RGD peptide ligands-design of new molecular conjugates for imaging and therapy of cancers. Anticancer Agents Med Chem 7(5):552–558

83. Wu Z, Li ZB, Chen K, Cai W, He L, Chin FT, Li F, Chen X (2007) micro PET of tumor integrin alphavbeta3 expression using 18F-labeled PEGylated tetrameric RGD peptide (18 F-FPRGD4). J Nucl Med 48(9):1536–1544

84. Lin H-Y, Davis FB, Gordinier JK, Martino LJ, Davis PJ (2001) Thyroid hormone induces activation of mitogen-activated protein kinase. Am J Physiol 276:C1014–C1024

85. Hercbergs AA, Goyal LK, Suh JH, Lee S, Reddy CA, Cohen BH, Stevens GH, Reddy SK, Peereboom DM, Elson PJ, Gupta MK, Barnett GH (2003) Propylthiouracil-induced chemical

hypothyroidism with high-dose tamoxifen prolongs survival in recurrent high grade glioma: a Phase I/II study. Anticancer Res 23:617–626

86. Cristofanilli M, Yamamura Y, Kau S-W, Bevers T, Strom S, Patangan M, Hsu L, Krishnamurthy S, Theriault RL, Hortobagyi GN (2005) Thyroid hormone and breast carcinoma. Primary hypothyroidism is associated with a reduced incidence of primary breast carcinoma. Cancer 103:1122–1128

87. Albert JM, Cao C, Geng L, Leavitt L, Hallahan DE, Lu B (2006) Integrin alpha v beta 3 antagonist Cilengitide enhances efficacy of radiotherapy in endothelial cell and non-small-cell lung cancer models. Int J Radiat Oncol Biol Phys 65(5):1536–1543

88. Monnier Y, Farmer P, Bieler G, Imaizumi N, Sengstag T, Alghisi GC, Stehle JC, Ciarloni L, Andrejevic-Blant S et al (2008) CYR61 and alphaVbeta5 integrin cooperate to promote invasion and metastasis of tumors growing in preirradiated stroma. Cancer Res 68(18):7323–7331

89. Lark MW, Stroup GB, Dodds RA et al (2001) Antagonism of the osteoclast vitronectin receptor with an orally active non peptide inhibitor prevents cancellous bone loss in the ovariectomized rat. J Bone Miner Res 16:319–327

90. Davis J, Warwick J, Totty N, Philp R, Helfich M, Horton M (1989) The osteoclast functional antigen, implicated in regulation of bone resorption, is biochemically related to the vitronectin receptor. J Cell Biol 109:1817–1826

91. Arap W, Pasqualin A, Ruoslahti E (1998) Cancer treatment by targeted drug delivery to tumor vasculature in a mouse model. Science 279:377–380

92. Raboisson P, Manthey CL, Chaikin M, Lattanze J, Crysler C, Leonard K, Pan W, Tomczuk BE, Marugan JJ (2006) Novel potent and selective alphavbeta3/alphavbeta5 integrin dual antagonists with reduced binding affinity for human serum albumin. Eur J Med Chem 41(7):847–861

93. Sipkins DA, Cheresh DA, Kazemi MR, Nevin LM, Bednarski MD, King CP (1998) Detection of tumor angiogenesis in vivo by αvβ3-targeted magnetic resonance imaging. Nat Med 4(5):623–626

94. Sivolapenko GB, Skarlos D, Pectasides D et al (1998) Imaging of metastatic melanoma utilising a technetium-99m labelled RGD-containing synthetic peptide. Eur J Nucl Med 25(10):1383–1389

95. Haubner R, Wester HJ, Reuning U et al (1999) Radiolabeled αvβ3 integrin antagonists: a new class of tracer for tumor targeting. J Nucl Med 40(6):1061–1071

96. Dijkgraaf I, Kruijtzer JA, Frielink C, Soede AC, Hilbers HW, Oyen WJ, Corstens FH, Liskamp RM, Boerman OC (2006) Synthesis and biological evaluation of potent alphavbeta3-integrin receptor antagonists. Nucl Med Biol 33(8):953–961

97. Chen X, Sievers E, Hou Y, Park R, Tohme M, Bart R, Bremner R, Bading JR, Conti PS (2005) Integrin αvβ3-targeted imaging of lung cancer. Neoplasia 7(3):271–279

98. Silvestri E, Schiavo L, Lombardi A, Goglia F (2005) Thyroid hormones as molecular determinants of thermogenesis. Acta Physiol Scand 184:265–283

99. Wrutniak-Cabello C, Casas F, Cabello G (2001) Thyroid hormone action in mitochondria. J Mol Endocrinol 26:67–77

100. Huang CJ, Geller HM, Green WL, Craelius W (1999) Acute effects of thyroid hormone analogs on sodium currents in neonatal rat myocytes. J Mol Cell Cardiol 31:881–893

101. Sakaguchi Y, Cui G, Sen L (1996) Acute effects of thyroid hormone on inward rectifier potassium channel currents in guinea pig ventricular myocytes. Endocrinology 137:4744–4751

102. Incerpi S, Luly P, De Vito P, Farias RN (1999) Short-term effects of thyroid hormones on the Na/H antiport in L-6 myoblasts: high molecular specificity for 3,5,3′-triiodo-L-thyronine. Endocrinology 140:683–689

103. Ashizawa K, Cheng S (1992) Regulation of thyroid hormone receptor-mediated transcription by a cytosol protein. Proc Natl Acad Sci USA 89:9277–9281

104. Vie MP, Evrfard C, Osty J, Breton-Gilet A, Blanchet P, Pomerance M, Rouget P, Francon J, Blondeau JP (1997) Purification, molecular cloning, and functional expression of the human nicotinamide-adenine dinucleotide phosphate-regulated thyroid hormone-binding protein. Mol Endocrinol 11:1728–1736

105. Bergh JJ, Lin H-Y, Lansing L, Mohamed SN, Davis FB, Mousa S, Davis PJ (2005) Integrin αvβ3 contains a cell surface receptor site for thyroid hormone that is linked to activation of mitogen-activated protein kinase and induction of angiogenesis. Endocrinology 146:2864–2871

106. Davis PJ, Davis FB, Cody V (2005) Membrane receptors mediating thyroid hormone action. Trends Endocrinol Metab 16:429–435

107. Mousa SA, O'Connor L, Davis FB, Davis PJ (2006) Proangiogenesis action of the thyroid hormone analog 3,5-diiodothyropropionic acid (DITPA) is initiated at the cell surface and is integrin-mediated. Endocrinology 147:1602–1607

108. D'Arezzo S, Incerpi S, Davis FB, Acconia F, Marino M, Farias RN, Davis PJ (2004) Rapid nongenomic effects of 3,5,3′-triiodo-L-thyronine on the intracellular pH of L-6 myoblasts are mediated by intracellular calcium mobilization and kinase pathways. Endocrinology 145:5694–5703

109. Davis FB, Mousa SA, O'Connor L, Mohamed S, Lin H-Y, Cao HJ, Davis P (2004) Proangiogenic action of thyroid hormone is fibroblast growth factor-dependent and is initiated at the cell surface. Circ Res 94:1500–1506

110. Mousa SA, O'Connor LJ, Bergh JJ, Davis FB, Scanlan TS, Davis PJ (2005) The proangiogenic action of thyroid hormone analogue GC-1 is initiated at an integrin. J Cardiovasc Pharmacol 46:356–360

111. Tang H-Y, Lin H-Y, Zhang S, Davis FB, Davis PJ (2004) Thyroid hormone causes mitogen-activated protein kinase-dependent phosphorylation of the nuclear estrogen receptor. Endocrinology 145:3265–3272

112. Davis FB, Tang H-Y, Shih A, Keating T, Lansing L, Hercbergs A, Fenstermaker RA, Mousa A, Mousa SA, Davis PJ, Lin H-Y (2006) Acting via a cell surface receptor, thyroid hormone is a growth factor for glioma cells. Cancer Res 66(14):7270–7275

113. Davis PJ, Davis FB, Mousa SA, Luidens MK, Lin HY (2011) Membrane receptor for thyroid hormone: physiologic and pharmacologic implications. Annu Rev Pharmacol Toxicol 51:99–115.

114. Mousa SA, Bergh JJ, Dier E, Rebbaa A, O'Connor LJ, Yalcin M, Aljada A, Dyskin E, Davis FB, Lin HY, Davis PJ (2008) Tetraiodothyroacetic acid, a small molecule integrin ligand, blocks angiogenesis induced by vascular endothelial growth factor and basic fibroblast growth factor. Angiogenesis 11(2):183–190

115. Xiong J-P, Stehle T, Diefenbach B, Zhang R, Dunker R, Scott DL, Joachimiak A, Goodman SL, Arnaout MA (2001) Crystal structure of the extracellular segment of integrin αvβ3. Science 294:339–345

116. Lin HY, Sun M, Tang HY, Lin C, Luidens MK, Mousa SA, Incerpi S, Drusano GL, Davis FB, Davis PJ (2009) L-Thyroxine vs. 3,5,3′-triiodo-L-thyronine and cell proliferation: activation of mitogen-activated protein kinase and phosphatidylinositol 3-kinase. Am J Physiol Cell Physiol 296(5):C980–C991.

117. Tang HY, Lin HY, Zhang S, Davis FB, Davis PJ (2004) Thyroid hormone causes mitogen-activated protein kinase-dependent phosphorylation of the nuclear estrogen receptor. Endocrinology 145(7):3265–3272

Chapter 12
Anti-angiogenesis Therapy as an Adjunct to Chemotherapy in Oncology

Shaker A. Mousa and Laila H. Anwar

Abstract On November 18th 2011, the FDA announced the anticipated removal of Avastin's (bevacizumab) indication for metastatic breast cancer based on the lack of evidence supporting its use for the improvement of overall survival. Avastin, much like other anti-angiogenic therapies, has been shown to slow but not prevent metastasis when given in combination with chemotherapy. An anti-angiogenic therapy that targets multiple pro-angiogenic factors is the focus of much research. In this chapter we present the anti-angiogenesis therapies available, obstacles to the prevention of tumor neovascularization, and finally delve into new therapies showing results that may change the future of oncology.

Introduction

In 2008, the estimated United States cancer prevalence count was 11,957,599 people [1]. In 2012, about 577,190 Americans are expected to die of cancer [2]. While cancer cases are still on the rise, mortality rates are gradually declining thanks to treatment advancements. Cancer is caused by both intrinsic and extrinsic factors. Intrinsic factors can be described as genetic mutations, hormones, and immune conditions. Extrinsic factors are smoking, infections and exposure to chemicals and radiation [2]. When the immune system is compromised, mutations (either inherited or caused by damage) can lead to tumor formation and to the start of a malignant process.

Angiogenesis is an innate process of the body that results in the development of new blood vessels, and is usually initiated in response to an increased demand for oxygen and nutrients by the cells of the body. The process is managed through pro-angiogenic and anti-angiogenic factors (Table 12.1).

S.A. Mousa (✉) • L.H. Anwar
The Pharmaceutical Research Institute at Albany College of Pharmacy
and Health Sciences, Rensselaer, NY, USA
e-mail: shaker.mousa@acphs.edu

S.A. Mousa and P.J. Davis (eds.), *Angiogenesis Modulations in Health and Disease:*   143
Practical Applications of Pro- and Anti-angiogenesis Targets,
DOI 10.1007/978-94-007-6467-5_12, © Springer Science+Business Media Dordrecht 2013

Table 12.1 Pro-angiogenic factors and anti-angiogenic factors that regulate angiogenesis

Pro-angiogenic factors	Anti-angiogenic factors
Vascular Endothelial Growth Factor (VEGF)	Angiostatin
Fibroblast Growth Factor (FGF)	Endostatin
Platelet Derived Growth Factor (PDGF)	Vasostatin
Epidermal Growth Factor (EGF)	Prolactin
Placental Growth Factor (PlGF)	Angiopoietin-2
Hypoxia inducible factor-1α (HIF-1α)	Interferon-α (IF-α)
Angiogenin	Interferon-γ (IF-γ)
Interleukin-8 (IL-8)	Interleukin-12 (IL-12)
Angiopoietin-1	Fibronectin
	Platelet factor-4

A tumor begins as a healthy cell that becomes malignant and multiplies. Tumors that are less than 1 mm in diameter attain their nourishment through the process of diffusion [3]. Once a tumor has reached a sufficient size (>2 mm diameter) it will require more than just diffusion for its supply and will need to create new capillary beds to assist in the process of development. Much like normal cells of the body, tumor cells use pro-angiogenic factors to increase oxygen and nutrient supplies through the formation of new vasculature branching from previous vessels in and around the tumor.

Angiogenesis is thought to be a major contributor to tumor growth and metastasis. Numerous anti-angiogenic therapies have been produced for use in combination with cytotoxic chemotherapy. Even though theory suggests that anti-angiogenic therapy should improve the outcomes of cancer patients, results of recent trials (AVADO [4], RiBBON-1 [5]) show that anti-angiogenic therapies may not improve overall survival. Therefore, in order to ensure proper inhibition of tumor metastasis there is a need to understand the mechanism of angiogenesis.

In this chapter, the need for and utility of anti-angiogenic treatment in cancer as an adjunct to chemotherapy will be assessed based on the physiology and development of tumor cells through angiogenesis.

Biomarkers in Angiogenesis

Biomarkers are molecules or genes that are monitored to track certain changes that occur in the body. Biomarkers are used to diagnose and track many diseases of the body, including cancer. Anti-angiogenic therapies must target a sufficient quantity of these biomarkers in order to adequately end the progress of angiogenesis. The need for certain anti-angiogenic therapies must also be reconsidered based on economic significance and cost versus benefit analyses for treatment in combination with chemotherapy versus chemotherapy alone. A brief discussion of relevant biomarkers follows.

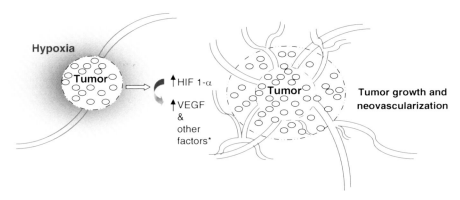

Fig. 12.1 Depiction of a tumor surrounded by a region of hypoxia. As a result of this condition, HIF-1α (hypoxia inducible factor-1α) is released and supports the release of pro-angiogenic factors such as VEGF (vascular endothelial growth factor) that initiate and promote the growth of new blood vessels. These blood vessels provide the tumor with oxygen and nutrients. *Other factors include EGF (epidermal growth factor), and IGF-2 (insulin-like growth factor-2)

Hypoxia Inducible Factor-1α

Angiogenesis is initiated by a decline in oxygen supply surrounding the newly formed tumor, also known as hypoxia. Hypoxia occurs due to metabolic demand (increased demand for O_2) produced by multiplying cells within the tumor. Hypoxia leads to the release of hypoxia inducible factor-1α (HIF-1α). The activity of HIF-1α promotes the progression of angiogenesis and metastases through the release of growth factors such as vascular endothelial growth factor (VEGF), EGF (epidermal growth factor), and IGF-2 (insulin-like growth factor-2) as depicted in Fig. 12.1. Increased levels of HIF-1α and VEGF have been linked with increased mortality in cancer [6–8].

Vascular Endothelial Growth Factor

VEGF, also known as vascular permeability factor, is responsible for the activation of the VEGF receptor located on the surface of vascular endothelial cells. The VEGF family consists of VEGF-A, VEGF-B, VEGF-C, VEGF-D, and PlGF (placental growth factor). There are three types of VEGF tyrosine kinase receptors on the surface of endothelial cells: VEGF receptor-1 (VEGFR-1), VEGF receptor-2 (VEGFR-2), and VEGF receptor-3 (VEGFR-3). Once the VEGF ligand binds to its receptor, a series of signals is set off to initiate angiogenesis, endothelial cell migration, and permeability. The VEGF-A factor binds to both VEGFR-1 and -2; VEGF-B and PlGF factors bind to VEGFR-1 only. Once VEGF binds to its receptor, the activated VEGF receptor will stimulate the AKT/mTOR signaling pathway through phosphoinositide 3-kinase (PI3K), leading to additional phosphorylation downstream and to the formation of mTOR complexes [9, 10].

Matrix Metalloproteinases

Matrix metalloproteinases (MMPs) are a group of enzymes that degrade the extracellular environment to assist in the process of new vessel formation, by allowing for endothelial cell migration to and around the tissue surrounding the tumor [11].

Current Anti-angiogenic Therapies

Angiogenesis is currently targeted with biologic and small molecule agents. The FDA has approved several agents such as bevacizumab (trade name Avastin®) and sorafenib (trade name Nexavar®), which have shown promising results when used in combination with chemotherapy. These therapies are only approved as adjuncts to chemotherapy and should never be used alone to treat cancer. Due to the large variation in biomarkers involved in the angiogenesis process, many anti-angiogenic therapies vary by the types of factor(s) that they target. The major targets of current therapy are VEGF and growth factor receptors. Other biomarkers such as HIF-1α and MMPs are major contributors to the angiogenic process but are yet to be included as targets in current therapy. A brief discussion of relevant therapies follows.

Bevacizumab Status in Oncology

In the E2100 trial published in 2007, paclitaxel versus paclitaxel plus bevacizumab for the treatment of metastatic breast cancer showed a significant increase in progression-free survival in the paclitaxel plus bevacizumab group (11.8 months) versus the paclitaxel group (5.9 months) [12]. The accelerated FDA approval for bevacizumab in metastatic breast cancer came in 2008 based on the results of the E2100 trial. The FDA requested that further trials be conducted to show that the addition of bevacizumab to chemotherapy provided an improvement in overall survival [13]. The AVADO three-arm trial published in 2010 compared two doses of bevacizumab plus docetaxel versus placebo plus docetaxel for the first line treatment of HER-2 negative metastatic breast cancer. Results of the trial showed a significant increase in progression-free survival for the higher dose bevacizumab-containing group, but the overall survival analysis revealed little difference between the three groups [4]. The RiBBON trials published in 2011 also showed a lack of change in overall survival with the addition of bevacizumab to standard chemotherapy [5]. Based on the results of the AVADO and RiBBON trials, the FDA insisted that the indication for metastatic breast cancer be revoked. On November 18, 2011 the FDA commissioner removed bevacizumab's indication for metastatic breast cancer since it was shown not to be cost-effective when combined with chemotherapy [14].

Monoclonal Antibodies / Recombinant Proteins

Bevacizumab is a humanized recombinant monoclonal IgG1 antibody and prevents the binding of VEGF to the VEGF receptors FLT1 and KDR on the surface of endo-thelial cells. Bevacizumab inhibits the biological activity of VEGF, which therefore inhibits angiogenesis [15]. Currently, bevacizumab is FDA-approved for first-line treatment of metastatic non-small cell lung cancer in combination with carboplatin and paclitaxel; first and second line treatment of metastatic colorectal cancer in combination with intravenous 5-fluorouracil-based chemotherapy; metastatic renal cell carcinoma in combination with interferon alfa; and as a single agent in the treat-ment of adult glioblastoma with progressive disease in post-chemotherapy patients.

VEGF-trap (aflibercept) is a recombinant fusion protein that binds VEGF-A and PlGF [16], inhibiting their binding to and activation of VEGF receptors. Aflibercept is currently FDA-approved for the treatment of patients with neovascular age-related macular degeneration [17]. In terms of its use in cancer, aflibercept has shown intermediate results and even poor results in the treatment of non-small cell lung cancer and pancreatic cancer. Lack of an FDA approval for cancer does not limit its potential use as an anti-angiogenic treatment for tumors. Aflibercept versus placebo has shown promising improvement in overall survival during an ongoing trial in combination with irinotecan and 5-FU in the treatment of patients with meta-static colorectal cancer after failure of an oxaliplatin-based regimen (VELOUR) [18] and has only shown positive results for its adjunct use in metastatic colon can-cer in phase III trials.

AMG 386 is an original peptide-Fc fusion protein that prevents the binding of angiopoietin-1 and angiopoietin-2 with Tie2 receptors [19]. This agent is showing potential in its use in combination with chemotherapy for the treatment of ovarian cancer. AMG 386 is undergoing a phase III clinical trial named 'Trinova-3: A study of AMG 386 or AMG 386 placebo in combination with paclitaxel and carboplatin to treat ovarian cancer' [20]. Long-term results are yet to be determined with this agent.

Ramucirumab is a 100 % humanized monoclonal antibody that prevents the binding of VEGF to VEGFR-2. This agent has shown antitumor and anti-angiogenic results during phase I and II trials conducted to treat renal cell carcinoma, hepato-cellular carcinoma, non-small cell lung cancer, and melanoma [21]. It is currently undergoing phase III trials. This agent differs from bevacizumab because bevaci-zumab only blocks the binding of VEGF-A, whereas ramucirumab blocks the bind-ing of all VEGF types to its receptor. Ramucirumab is also very specific with a high affinity for the VEGF receptor epitope, unlike other tyrosine kinase inhibitors that bind to non-tumor areas, leading to a higher risk of toxicity [21, 22].

DI17E6 is a monoclonal antibody to human αv integrins. It effectively reduced angiogenesis in prostate cancer by preventing cell adhesion and migration. DI17E6 was used to target the integrin receptor expressed on the surface of tumor cells for the delivery of chemotherapy. In the Wagner et al. study, DI17E6 was bonded with a doxorubicin nanoparticle. When compared with conventional drug delivery, the DI17E6-linked cytotoxic nanoparticle showed an increased level of activity [23].

There are numerous monoclonal antibodies still in early phase II trials, two of which, MetMAb and Pegdinetanib, have shown efficacy in preventing angiogenesis in non-small cell lung cancer and glioblastomas, respectively.

Tyrosine Kinase Inhibitors

Sunitinib (trade name Sutent®), a multi-receptor tyrosine kinase inhibitor, is a small molecule that binds to and stops the activation of receptors PDGFRα, PDGFRβ, VEGFR-1, VEGFR-2, VEGFR-3, Kit, FLT3, CSF-1R, and RET. The FDA has approved sunitinib's use in the treatment of gastrointestinal stromal tumors, renal cell carcinoma, and in advanced pancreatic neuroendocrine tumors. Unlike biologic agents that require intravenous administration, sunitinib is taken orally in a capsule form [24, 25].

Sorafenib is a multi-receptor tyrosine kinase inhibitor that is FDA-indicated for the treatment of unresectable hepatocellular carcinoma [26] and advanced renal cell carcinoma. This agent acts intracellularly on receptors CRAF, BRAF, and mutant BRAF, and extracellularly on receptors Kit, FLT3, RET, VEGFR-1, VEGFR-2, VEGFR-3, and PDGFRβ [27]. In the Escudier et al. study comparing sorafenib to placebo in post-treated patients with renal cell carcinoma, overall survival was not found to be statistically significant, but there was an overall benefit noted with its administration [28].

Vandetanib (trade name Caprelsa®) is a tyrosine kinase inhibitor that binds to the EGF receptors, VEGF receptors, RET, protein tyrosine kinase 6, Tie2, EPH receptor kinases, and Src tyrosine kinases. The overall result of vandetanib binding is a decrease in endothelial cell migration, proliferation, survival, and neovascularization. Vandetanib is currently FDA-indicated for the treatment of symptomatic or progressive medullary thyroid cancer in patients with unresectable, locally advanced or metastatic disease [29]. In a phase II, open label study of 30 patients with diagnosed medullary thyroid cancer, treatment with 300 mg vandetanib demonstrated a confirmed partial antitumor response in 20 % of patients and stabilized disease for 53 % of patients at ≥24 weeks [30].

Pazopanib (trade name Votriant®) is a multi-tyrosine kinase inhibitor of receptors VEGFR-1, VEGFR-2, VEGFR-3, PDGFRα, PDGFRβ, FGFR1, FGFR3, Kit, IL-2, and c-FMS [31]. Pazopanib is FDA-indicated for the treatment of renal cell carcinoma based on the results of a phase III trial testing pazopanib versus placebo treatment in adult patients with measurable, locally advanced, and/or metastatic renal cell carcinoma. Patients were randomized in a 2:1 ratio to receive 800 mg of pazopanib once daily or placebo and were treated until disease progression, death, toxicity, or withdrawal. Of the 435 patients enrolled, 290 patients were assigned to take pazopanib and 145 patients were assigned to take placebo. Progression-free survival was significantly prolonged in the overall study population treated with pazopanib versus placebo by 9.2 months and 4.2 months, respectively [32, 33].

Some newer small molecule anti-angiogenic therapies that are showing positive results in clinical trials but are not yet approved are tivozanib (a VEGF receptor inhibitor), axitinib (a multikinase inhibitor), motesanib (a multikinase inhibitor), intedanib (a VEGFR-2/PDGFR/FGFR inhibitor), and brivanib (a VEGFR-2/FGFR1 inhibitor).

Obstacles to Current Anti-angiogenic Therapy

The current anti-angiogenic therapies show slight effectiveness in halting angiogenesis but do not stop it completely. Even with the FDA-approved agents such as bevacizumab, sorafenib, and sunitinib, angiogenesis continues to lead to the metastasis and mortality associated with cancer.

The current understanding is that angiogenesis is initiated and carried out by the activation of several biomarkers. It has been shown that VEGF inhibition alone is not enough to stop the process of neovascularization. There are multi-receptor tyrosine kinases such as sunitinib and pazopanib that target more than one receptor in the angiogenesis process but are also failing to prevent the metastatic process. In addition, tyrosine kinase inhibitors are known to cause several undesirable side effects due to the lack of specificity in binding sites. A need for an agent that targets multiple biomarkers and receptors in the angiogenic process with a high specificity for tumor binding sites is needed and is a current focus of research.

Vascular Abnormalities

VEGF is known to reduce the interstitial pressure surrounding the tumor by reducing the fluid surrounding leaky vessels, and it also reduces ascites fluids. It has also been suggested that VEGF levels are higher in malignant pleural effusions surrounding tumors in comparison to nonmalignant pleural effusions [34]. When the amount of fluid surrounding the tumor is normalized, the oxygen and blood flow to the tumor will improve and thus will improve chemotherapy delivery to the tumor. The level of VEGF surrounding endothelial cells will also be reduced [35].

Razoxane and dexrazoxane are other agents that have shown anti-angiogenic properties. They are thought to reduce leaky vessels within the tumors and to reduce interstitial pressure surrounding the tumor site. In a study conducted to determine the anti-angiogenic effects of dexrazoxane, it was determined that the agent causes an up-regulation of THBS-1, an endogenous inhibitor of neovascularization [36].

Interstitial Pressure

As a tumor progresses, interstitial fluid pressure surrounding the tumor increases due to the increased permeability of vessels and the lack of lymphatic drainage. This increase in interstitial pressure is associated with hypoxia and harbors pro-angiogenic factors at the outer rim of the tumor. Based on the results of the Phipps and Kohandel mathematical model, interstitial fluid pressure gradients hypothetically affect the inhibition and stimulation of angiogenesis: "Angiogenic behaviors were suppressed when closer to the core of the tumor and maximal angiogenic stimulation was detected towards the [outer] rim of the tumor" [37].

Angiogenesis and Thrombosis

Platelet activation and fibrin formation are mechanisms by which tumor cells promote angiogenesis and metastasis. Coagulation proteases consisting of platelet and fibrin form on the surface of tumor cells. This process promotes tumor cell conjunction into a small thrombus that can lodge into the vasculature [38]. A fibrin matrix is eventually formed surrounding the tumor. It has been noted that heparin is effective in inhibiting the proliferation of endothelial cells and hindering angiogenesis via structural changes to fibrin networks surrounding tumors [39]. Low molecular weight heparin (LMWH) is superior to unfractionated heparin (UFH): in a study comparing the two anticoagulants it was found that UFH enhanced the fibrin matrix and had a pro-angiogenic outcome whereas LMWH had an anti-angiogenic effect on the tumor [39]. It is also beneficial to add "thrombosis prophylaxis" to anti-angiogenic therapy because of an increased risk of thromboembolism and clot formation with the administration of these agents [40].

Novel Therapies

Angiogenesis has been treated with agents that inhibit one or a few of the many ways that tumors initiate angiogenesis and progressively become metastatic. The recent revocation of bevacizumab's indication for metastatic breast cancer by the FDA is an example of failure with such a narrow treatment goal. The need for therapies that target multiple receptors and biomarkers is a trend that has been producing a number of new agents that are showing promising results in clinical trials.

Tetrac, a derivate of thyroid hormone, and tetrac formulated as nanoparticles (nanotetrac), have both shown anti-proliferative and anti-angiogenic action against cancer cells in clinical trials. T4 (thyroxine) and T3 (triiodothyronine) hormones bind to thyroid hormone receptors located on integrin $\alpha v\beta 3$ membrane surface proteins, which leads to pro-angiogenic effects [41]. Tetrac and nanotetrac were tested *in vitro* and in xenografts for the treatment of lung cancer. It was observed that tetrac and nanotetrac successfully inhibited the binding of T4 and T3 to the receptor located on the tumor cell's surface and prevented hormone activity on the tumor cell. Nanotetrac proved to be superior to unmodified tetrac because it showed greater effects on malignant cell gene expression and was more potent [42]. Further studies with tetrac and nanotetrac are needed to determine the long-term effects and success of tetrac treatment.

Thalidomide, an anti-inflammatory, anti-angiogenic and immune-modulating agent, has been used to treat conditions such as cutaneous manifestations of erythema nodosum leprosum and in combination with dexamethasone to treat newly diagnosed multiple myeloma. This agent reduces the levels of tumor necrosis factor-alpha, increases the number of natural killer cells and IL-2, and decreases the proliferation of endothelial cells [43]. Thalidomide reduces angiogenesis by altering the tumor's microenvironment via decreasing the secretion of pro-angiogenic factors such as VEGF and IL-6. Thalidomide has been tested to treat both blood and solid tumors. The success found in thalidomide treatment of multiple myeloma has not been matched by trials testing thalidomide's use in hepatocellular and colorectal cancer solid tumors [44, 45].

Lenalidomide (trade name Revlamid ®), an immune-modulating drug, acts much like thalidomide and is indicated for the treatment of patients with refractory multiple myeloma when used in combination with dexamethasone and for the treatment of myelodysplastic syndromes [46]. In a study comparing lenalidomide and dexamethasone versus placebo and dexamethasone for the treatment of relapsed multiple myeloma, complete or partial responses were observed in 61 % of the lenalidomide group (108 out of 177 patients) and 19.9 % of the dexamethasone-only group (35 out of 176 patients) [47]. This study proved that the administration of lenalidomide in combination with dexamethasone is statistically superior to the administration of dexamethasone alone for multiple myeloma treatment.

The idea of linking chemotherapy agents to anti-VEGF receptor antibodies has also been employed and tested, with promising outcomes. In a study by Wicki et al. an anti-VEGFR-2 antibody was linked with doxorubicin to inhibit tumor growth directly. It was found that the doxorubicin-linked VEGFR antibody provided better selectivity and suppression of the tumor vasculature than doxorubicin or VEGFR antibody alone [48]. The practice of using a highly expressed receptor on the surface of tumor cells to deliver chemotherapy allows for greater binding specificity and decreases the effects of chemotherapy-related adverse effects.

Conclusions

There is a current need for further understanding of biomarkers, initiators of neovascularization, and sustainers of tumor-related angiogenesis. When the process of tumor angiogenesis is understood, a therapy can be targeted and tailored to the tumor cell, making the results of treatment more efficient and beneficial. This does not imply that current therapies such as monoclonal antibodies and tyrosine kinase inhibitors will become futile. These therapies can be given in combination to provide a multi-targeted approach to therapy. They can also be used to deliver chemotherapy directly to tumor cells. Tumor barriers such as exudated fluid and fibrin matrices must be removed or permeated in order for anti-angiogenics and chemotherapy to reach the tumor cells. The use of non-cytotoxic agents such as low molecular weight heparin for the reduction of fibrin clots and vascular normalizing agents such as razoxane and dexrazoxane to reduce interstitial fluid pressure provide a synergistic effect when administered with anti-angiogenic treatment and chemotherapy.

The future of oncology may be shifting towards a multi-treatment approach involving both cytotoxic and non-cytotoxic therapies. Chemotherapy may eventually be administered more efficiently through the use of nanotechnology and targeted antibodies. The modification of standard chemotherapy administration will not only reduce the risk of systemic effects but also provide better survival outcomes for the cancer patient.

References

1. SEER Cancer Statistics Review (1975–2008) National Cancer Institute. http://seer.cancer.gov/ csr/1975_2008/, based on November 2010 SEER data submission, posted to the SEER web site, 2011. Accessed 12 Mar 2013
2. American Cancer Society Cancer Facts and Figures (2012) American Cancer Society. http:// www.cancer.org/research/cancerfactsfigures/cancerfactsfigures/cancer-facts-figures-2012. Accessed 12 Mar 2013
3. Singhal S, Vachani A, Antin-Ozerkis D, Kaiser LR, Albelda SM (2005) Prognostic implications of cell cycle, apoptosis, and angiogenesis biomarkers in non-small cell lung cancer: a review. Clin Cancer Res 11(11):3974–3986
4. Miles DW, Chan A, Dirix LY, Cortes J, Pivot X, Tomczak P, Delozier T, Sohn JH, Provencher L, Puglisi F, Harbeck N, Steger GG, Schneeweiss A, Wardley AM, Chlistalla A, Romieu G (2010) Phase III study of bevacizumab plus docetaxel compared with placebo plus docetaxel for the first-line treatment of human epidermal growth factor receptor 2-negative metastatic breast cancer. J Clin Oncol 28(20):3239–3247
5. Robert NJ, Dieras V, Glaspy J, Brufsky AM, Bondarenko I, Lipatov ON, Perez EA, Yardley DA, Chan SY, Zhou X, Phan SC, O'Shaughnessy J (2011) RIBBON-1: randomized, double-blind, placebo-controlled, phase III trial of chemotherapy with or without bevacizumab for first-line treatment of human epidermal growth factor receptor 2-negative, locally recurrent or metastatic breast cancer. J Clin Oncol 29(10):1252–1260
6. Schwab LP, Peacock DL, Majumdar D, Ingels JF, Jensen LC, Smith KD, Cushing RC, Seagroves TN (2012) Hypoxia inducible factor-1 alpha promotes primary tumor growth and tumor-initiating cell activity in breast cancer. Breast Cancer Res 14(1):R6
7. Mooring SR, Jin H, Devi NS, Jabbar AA, Kaluz S, Liu Y, Van Meir EG, Wang B (2011) Design and synthesis of novel small-molecule inhibitors of the hypoxia inducible factor pathway. J Med Chem 54(24):8471–8489
8. Birner P, Schindl M, Obermair A, Plank C (2000) Overexpression of hypoxia-inducible factor 1alpha is a marker for an unfavorable prognosis in early-stage invasive cervical cancer. Cancer Res 60:4693–4696
9. Trinh XB, Tjalma WA, Vermeulen PB, Van den Eynden G, Van der Auwera I, Van Laere SJ, Helleman J, Berns EM, Dirix LY, van Dam PA (2009) The VEGF pathway and the AKT/ mTOR/p70S6K1 signalling pathway in human epithelial ovarian cancer. Br J Cancer 100(6):971–978

10. Karar J, Maity A (2011) PI3K/AKT/mTOR pathway in angiogenesis. Front Mol Neurosci 4:51
11. Rundhaug JE (2005) Matrix metalloproteinases and angiogenesis. J Cell Mol Med 9(2):267–285
12. Miller K, Wang M, Gralow J, Dickler M, Cobleigh M, Perez EA, Shenkier T, Cella D, Davidson NE (2007) Paclitaxel plus bevacizumab versus paclitaxel alone for metastatic breast cancer. N Engl J Med 357(26):2666–2676
13. FDA Begins Process to Remove Breast Cancer Indication From Avastin Label (2010) http://www.fda.gov/NewsEvents/Newsroom/PressAnnouncements/2010/ucm237172.htm. Accessed 12 Mar 2013
14. Montero AJ, Avancha K, Gluck S, Lopes G (2012) A cost-benefit analysis of bevacizumab in combination with paclitaxel in the first-line treatment of patients with metastatic breast cancer. Breast Cancer Res Treat 132(2):747–751
15. Genentech Inc (2013) Avastin prescribing information. http://www.gene.com/download/pdf/avastin_prescribing.pdf. Accessed 12 Mar 2013
16. Hsu JY, Wakelee HA (2009) Monoclonal antibodies targeting vascular endothelial growth factor: current status and future challenges in cancer therapy. BioDrugs 23(5):289–304
17. Regeneron Pharmaceuticals Inc (2013) Eylea prescribing information. http://www.regeneron.com/Eylea/eylea-fpi.pdf. Accessed 12 Mar 2013
18. Sanofi-Aventis (On going phase III trial) (2013) A multinational, randomized, double-blind study, comparing the efficacy of aflibercept once every 2 weeks versus placebo in patients with metastatic colorectal cancer (MCRC) treated with irinotecan/5-FU combination (FOLFIRI) after failure of an oxaliplatin based Regimen. http://clinicaltrials.gov/ct2/show/NCT00561470?term=velour&rank=1. Accessed 12 Mar 2013
19. Herbst RS, Hong D, Chap L, Kurzrock R, Jackson E, Silverman JM, Rasmussen E, Sun YN, Zhong D, Hwang YC, Evelhoch JL, Oliner JD, Le N, Rosen LS (2009) Safety, pharmacokinetics, and antitumor activity of AMG 386, a selective angiopoietin inhibitor, in adult patients with advanced solid tumors. J Clin Oncol 27(21):3557–3565
20. Amgen (On going phase III trial) (2013) A phase 3 randomized, double-blind, placebo-controlled, multicenter study of AMG 386 with paclitaxel and carboplatin as first-line treatment of subjects with FIGO stage III-IV epithelial ovarian, primary peritoneal or fallopian tube cancers. http://clinicaltrials.gov/ct2/show/NCT01493505?term=trinova-3&rank=1. Accessed 12 Mar 2013
21. Spratlin JL, Cohen RB, Eadens M, Gore L, Camidge DR, Diab S, Leong S, O'Bryant C, Chow LQ, Serkova NJ, Meropol NJ, Lewis NL, Chiorean EG, Fox F, Youssoufian H, Rowinsky EK, Eckhardt SG (2010) Phase I pharmacologic and biologic study of ramucirumab (IMC-1121B), a fully human immunoglobulin G1 monoclonal antibody targeting the vascular endothelial growth factor receptor-2. J Clin Oncol 28(5):780–787
22. Yu L, Liang XH, Ferrara N (2011) Comparing protein VEGF inhibitors: in vitro biological studies. Biochem Biophys Res Commun 408(2):276–281
23. Wagner S, Rothweiler F, Anhorn MG, Sauer D, Riemann I, Weiss EC, Katsen-Globa A, Michaelis M, Cinatl J Jr, Schwartz D, Kreuter J, von Briesen H, Langer K (2010) Enhanced drug targeting by attachment of an anti alphav integrin antibody to doxorubicin loaded human serum albumin nanoparticles. Biomaterials 31(8):2388–2398
24. Pfizer Labs Inc (2012) Sutent prescribing information. http://labeling.pfizer.com/ShowLabeling.aspx?id=607. Accessed 12 Mar 2013
25. Shirao K, Nishida T, Doi T, Komatsu Y, Muro K, Li Y, Ueda E, Ohtsu A (2010) Phase I/II study of sunitinib malate in Japanese patients with gastrointestinal stromal tumor after failure of prior treatment with imatinib mesylate. Invest New Drugs 28(6):866–875
26. Llovet JM, Ricci S, Mazzaferro V, Hilgard P, Gane E, Blanc JF, de Oliveira AC, Santoro A, Raoul JL, Forner A, Schwartz M, Porta C, Zeuzem S, Bolondi L, Greten TF, Galle PR, Seitz JF, Borbath I, Haussinger D, Giannaris T, Shan M, Moscovici M, Voliotis D, Bruix J, Group SIS (2008) Sorafenib in advanced hepatocellular carcinoma. N Engl J Med 359(4):378–390
27. Bayer Healthcare Pharmceuticals Inc., Onxy Pharmaceuticals (2012) Nexavar prescribing information. http://www.nexavar-us.com/scripts/pages/en/patient/index.php. Accessed 12 Mar 2013

28. Escudier B, Eisen T, Stadler WM, Szczylik C, Oudard S, Staehler M, Negrier S, Chevreau C, Desai AA, Rolland F, Demkow T, Hutson TE, Gore M, Anderson S, Hofilena G, Shan M, Pena C, Lathia C, Bukowski RM (2009) Sorafenib for treatment of renal cell carcinoma: final efficacy and safety results of the phase III treatment approaches in renal cancer global evaluation trial. J Clin Oncol 27(20):3312–3318
29. Astra Zeneca Pharmaceuticals (2011) Caprelsa prescriber information. http://www1.astrazeneca-us.com/pi/caprelsa.pdf. Accessed 12 Mar 2013
30. Wells SA Jr, Gosnell JE, Gagel RF, Moley J, Pfister D, Sosa JA, Skinner M, Krebs A, Vasselli J, Schlumberger M (2010) Vandetanib for the treatment of patients with locally advanced or metastatic hereditary medullary thyroid cancer. J Clin Oncol 28(5):767–772
31. GlaxoSmithKline (2009) Votrient prescribing information. http://www.accessdata.fda.gov/drugsatfda_docs/label/2009/022465lbl.pdf. Accessed 12 Mar 2013
32. Sternberg CN, Davis ID, Mardiak J, Szczylik C, Lee E, Wagstaff J, Barrios CH, Salman P, Gladkov OA, Kavina A, Zarba JJ, Chen M, McCann L, Pandite L, Roychowdhury DF, Hawkins RE (2010) Pazopanib in locally advanced or metastatic renal cell carcinoma: results of a randomized phase III trial. J Clin Oncol 28(6):1061–1068
33. Gril B, Palmieri D, Qian Y, Anwar T, Ileva L, Bernardo M, Choyke P, Liewehr DJ, Steinberg SM, Steeg PS (2011) The B-Raf status of tumor cells may be a significant determinant of both antitumor and anti-angiogenic effects of pazopanib in xenograft tumor models. PLoS One 6(10):e25625
34. Masood R, Cai J, Zheng T, Smith DL, Hinton DR, Gill PS (2001) Vascular endothelial growth factor (VEGF) is an autocrine growth factor for VEGF receptor-positive human tumors. Blood 98(6):1904–1913
35. Carmeliet P, Moons L, Luttun A, Vincenti V, Compernolle V, De Mol M, Wu Y, Bono F, Devy L, Beck H, Scholz D, Acker T, DiPalma T, Dewerchin M, Noel A, Stalmans I, Barra A, Blacher S, Vandendriessche T, Ponten A, Eriksson U, Plate KH, Foidart JM, Schaper W, Charnock-Jones DS, Hicklin DJ, Herbert JM, Collen D, Persico MG (2001) Synergism between vascular endothelial growth factor and placental growth factor contributes to angiogenesis and plasma extravasation in pathological conditions. Nat Med 7(5):575–583
36. Maloney SL, Sullivan DC, Suchting S, Herbert JM, Rabai EM, Nagy Z, Barker J, Sundar S, Bicknell R (2009) Induction of thrombospondin-1 partially mediates the anti-angiogenic activity of dexrazoxane. Br J Cancer 101(6):957–966
37. Phipps C, Kohandel M (2011) Mathematical model of the effect of interstitial fluid pressure on angiogenic behavior in solid tumors. Comput Math Methods Med 2011:843765
38. Ruf W, Disse J, Carneiro-Lobo TC, Yokota N, Schaffner F (2011) Tissue factor and cell signalling in cancer progression and thrombosis. J Thromb Haemost 9(Suppl 1):306–315
39. Collen A, Smorenburg SM, Peters E, Lupu F, Koolwijk P, Van Noorden C, van Hinsbergh VW (2000) Unfractionated and low molecular weight heparin affect fibrin structure and angiogenesis in vitro. Cancer Res 60(21):6196–6200
40. Elice F, Rodeghiero F, Falanga A, Rickles FR (2009) Thrombosis associated with angiogenesis inhibitors. Best Pract Res Clin Hematol 22(1):115–128
41. Davis PJ, Davis FB, Mousa SA (2009) Thyroid hormone-induced angiogenesis. Curr Cardiol Rev 5(1):12–16
42. Mousa SA, Yalcin M, Bharali DJ, Meng R, Tang HY, Lin HY, Davis FB, Davis PJ (2012) Tetraiodothyroacetic acid and its nanoformulation inhibit thyroid hormone stimulation of non-small cell lung cancer cells in vitro and its growth in xenografts. Lung Cancer 76(1):39–45
43. Celgene Corporation (1998–2010) Thalidomide prescribing information. http://www.thalomid.com/. Accessed 8 April 2013
44. Lin AY, Brophy N, Fisher GA, So S, Biggs C, Yock TI, Levitt L (2005) Phase II study of thalidomide in patients with unresectable hepatocellular carcinoma. Cancer 103(1):119–125
45. McCollum AD, Wu B, Clark JW, Kulke MH, Enzinger PC, Ryan DP, Earle CC, Michelini A, Fuchs CS (2006) The combination of capecitabine and thalidomide in previously treated, refractory metastatic colorectal cancer. Am J Clin Oncol 29(1):40–44

46. Celgene Corporation (2013) Revlimid prescribing information. http://www.revlimid.com/. Accessed 12 Mar 2013
47. Weber DM, Chen C, Niesvizky R, Wang M, Belch A, Stadtmauer EA, Siegel D, Borrello I, Rajkumar SV, Chanan-Khan AA, Lonial S, Yu Z, Patin J, Olesnyckyj M, Zeldis JB, Knight RD, Multiple Myeloma Study Investigators (2007) Lenalidomide plus dexamethasone for relapsed multiple myeloma in North America. N Engl J Med 357(21):2133–2142
48. Wicki A, Rochlitz C, Orleth A, Ritschard R, Albrecht I, Herrmann R, Christofori G, Mamot C (2012) Targeting tumor-associated endothelial cells: anti-VEGFR2 immunoliposomes mediate tumor vessel disruption and inhibit tumor growth. Clin Cancer Res 18(2):454–464

Chapter 13
Anti-VEGF Strategies in Ocular Angiogenesis-mediated Disorders, with Special Emphasis on Age-related Macular Degeneration

Shaker A. Mousa

Abstract Pathological angiogenesis in the eye including exudative age-related macular degeneration (AMD), proliferative diabetic retinopathy, diabetic macular edema, neovascular glaucoma, and corneal neovascularization (trachoma) underlies the major causes of blindness in both developed and developing nations. Additionally, increased rates of angiogenesis are associated with several other disease states including cancer, psoriasis, rheumatoid arthritis and other vascular-associated disorders. Vascular endothelial growth factor (VEGF) and its receptors play an important role in the modulation of angiogenesis and have been implicated in the pathology of a number of conditions, including AMD, diabetic retinopathy, and cancer. AMD is a progressive disease of the macula and the third major cause of blindness worldwide. If not treated appropriately, AMD might progress to the second eye. Until recently, the treatment options for AMD were limited, with photodynamic therapy the mainstay treatment, which is effective at slowing disease progression but rarely results in improved vision. There are currently three approved anti-angiogenesis biologic therapies for ophthalmic diseases: an anti-VEGF aptamer (pegaptanib, Macugen®), a Fab fragment of a monoclonal antibody directed against VEGF-A (ranibizumab, Lucentis®), and VEGF trap (aflibercept, Eylea®). Several therapies have been and are now being developed for neovascular AMD, with the goal of inhibiting VEGF. At present, established therapies have met with great success in reducing the vision loss associated with neovascular AMD, whereas those still investigational in nature offer the potential for further advances. In AMD patients these therapies slow the rate of vision loss and in some cases increase visual acuity. Although these therapies are a milestone in the treatment of these disease states, several concerns need to be addressed before their impact can be fully understood.

S.A. Mousa (✉)
The Pharmaceutical Research Institute at Albany College of Pharmacy
and Health Sciences, Rensselaer, NY, USA
e-mail: shaker.mousa@acphs.edu

S.A. Mousa and P.J. Davis (eds.), *Angiogenesis Modulations in Health and Disease:*   157
Practical Applications of Pro- and Anti-angiogenesis Targets,
DOI 10.1007/978-94-007-6467-5_13, © Springer Science+Business Media Dordrecht 2013

Introduction

Angiogenesis is a term used to describe the formation of new blood vessels from the pre-existing vasculature. This process is critical for several normal physiological functions including the development of embryos, wound-healing, the female reproductive cycle and collateral vascular generation in the myocardium. However, aberrant angiogenesis has been implicated in the progression of several disease states including cancer, macular degeneration, diabetic retinopathy, rheumatoid arthritis and psoriasis.

Under normal physiological conditions, the process of angiogenesis is well controlled, and a perfect balance of endogenous pro-angiogenesis growth factors (positive regulator) and suppressors (negative regulator) exists. When angiogenic growth factors outnumber angiogenesis inhibitors, the balance shifts in favor of accelerated angiogenesis; this has been termed the "angiogenic switch" [1]. Rigorous research in the field of angiogenesis has led to the identification of many regulators involved in angiogenesis. Angiogenesis is driven by the production of pro-angiogenic growth factors including vascular endothelial growth factor (VEGF), basic fibroblast growth factor (bFGF), interleukin 8 (IL-8), placental like growth factor (PlGF), transforming growth factor β (TGF-β), angiopoietin, platelet-derived endothelial growth factor (PDEGF), pleiotrophin, and several others [2]. In addition, angiogenesis can be caused by a deficiency in endogenous angiogenesis inhibitors including angiostatin, canstatin, endostatin, various glycosaminoglycan, interferon α, β, χ (INF α, β, χ), thrombospondin and others [3].

Although angiogenesis is not understood in its entirety, the roles of many of its regulators and the fundamental steps that result in angiogenesis have been well documented. Initially, vascular endothelial cells (ECs) are activated by pro-angiogenesis growth factors, which cause ECs to release proteases that degrade the basement membrane, allowing ECs to escape from the original vessel walls, proliferate, and extend toward the source of the angiogenic stimulus using integrin and extracellular matrix proteins to cause cell adhesion [1, 3].

Pathological Angiogenesis

Cancer research has shown that due to a lack of oxygen and other essential nutrients, tumor growth is limited to 1–2 mm, and in order to grow beyond this size tumor cells must promote angiogenesis by secreting various pro-angiogenesis factors [3, 4]. Tumor angiogenesis not only allows tumor growth, but also increases the rate of metastasis. Vessels formed by uncontrolled and unregulated angiogenesis supporting the tumor are drastically different from those of the normal vasculature and are characterized by unstructured blood vessels, hypoxia, and increased interstitial pressure. These irregularities may hinder the ability of chemotherapeutic agents to achieve the desired effective levels within tumor.

Age-related macular degeneration (AMD) is a disease with complex pathology, which could be presented in either the dry (geographic atrophy) or the wet (choroidal neovascularization) form, and in some cases the dry form leads to the wet form. The dry form represents the majority of the AMD cases as opposed to the wet form. However, the wet form leads to progressive vision loss associated with major social and economic impact for the patient [5]. The role of VEGF in the accelerated angiogenesis process in choroidal angiogenesis has been documented, leading the assumption for the use of various anti-VEGF strategies [6]. The main purpose of this chapter is to summarize VEGF's physiological role (especially within the eye), the role in the development of AMD, and to understand and foresee both the benefits and potential side effects of the anti-VEGF-based therapy.

While the wet form of AMD can be managed using anti-angiogenesis strategies such as anti-VEGF [6], the dry form of advanced AMD results from atrophy of the retinal pigment epithelial layer and has no treatment option at this stage.

Research also shows that angiogenesis accompanies the progression of chronic inflammation. It has been demonstrated that VEGF is overexpressed in a number of pro-inflammatory conditions including psoriasis and rheumatoid arthritis [7, 8]. Thus, VEGF is an attractive target for the treatment of these diseases keeping in mind the redundancy in the pro-angiogenesis pathway and the potential for acquired resistance.

Anti-angiogenesis Therapies

A wide range of therapies designed to inhibit pathological angiogenesis have been developed and many more are underway. Angiogenesis inhibitors have typically been divided into two categories, either a direct strategy targeting ECs or an indirect strategy targeting pro-angiogenesis growth factors or their receptors. Direct targeting of ECs versus the case of a single pro-angiogenesis factor such as VEGF was thought to be a better target for therapy because it is relatively more genetically stable than cancer cells. It is postulated that this stability reduces the likelihood of rapid mutation and acquired drug resistance [9]. Recent studies suggest, however, that genetic anomalies are present in tumor ECs and may be able to confer drug resistance [10]. Interestingly, it has also been suggested that traditional therapies, such as radiation therapy, may actually work in part by targeting the genomically stable ECs because these ECs are still proliferating at a higher than normal rate [11].

Indirect inhibition of angiogenesis can be further divided into two categories, either amplifying the effects of angiogenesis inhibitors and the activation of their pathways or by inhibiting the activation of pro-angiogenesis pathways. Currently, these therapies have employed a multitude of targets including many angiogenic regulators and their receptors. One example is a therapy designed to target TGF. A clinical trial investigating the use of the transforming growth factor (TGF)-β antisense vaccine belagenpumatucel-L (Lucanix®) in patients with non-small cell lung cancer

(NSCLC) demonstrated favorable outcome as compared to historical control, with no observed adverse event [12, 13]. Another therapy being explored targets TGF and employs the use of a soluble TGF-ß receptor (sTGF-ßR) that specifically inhibits TGF-ß1 and TGF-ß3 [14].

VEGF

VEGF is a member of a family of dimeric glycoproteins that belong to the platelet-derived growth factor (PDGF) family of growth factors. While VEGF, also known as VEGF-A, is the most comprehensively studied member of the family, others include VEGF-B, VEGF-C, VEGF-D, and PlGF [15, 16]. VEGF-A has several isoforms (VEGF$_{121}$, VEGF$_{121}$b, VEGF$_{145}$, VEGF$_{165}$, VEGF$_{189}$, VEGF$_{206}$) resulting from alternative splicing of which VEGF$_{145}$ is most the most abundant isoform [17]. All VEGF ligands bind to tyrosine kinase receptors, causing the receptors to dimerize and phosphorylate [18]. Upon binding to its receptor, VEGF initiates a cascade of signaling events that begins with auto-phosphorylation of both receptor kinases, followed by activation of numerous downstream proteins including phospholipase Cλ, PI3K, GAP, Ras, MAPK and others [19]. VEGF receptor-2 (VEGFR-2) has a higher affinity for VEGF, and one of its biological activities includes the potentiation of angiogenesis [19]. The function of VEGFR-1 is less well defined, but seems to include recruitment of monocyte [19]. In contrast, VEGF-C and VEGF-D bind to a different receptor,VEGFR-3, which mediates lymphangiogenesis [16]. The biological activities of VEGF have also been well documented, and because of its vascular permeability characteristic, it was also named as a vascular permeability factor [20] It has also been shown to promote the growth, migration, and proliferation of ECs [20, 21]. In addition it induces vasodilatation and enhances EC survival [20, 21]. These biological activities occur in few physiological processes outside wound-healing and ovulation, making VEGF an attractive target for therapy.

VEGF Role in AMD

VEGF expression and its regulation were studied in retinal pigment epithelial (RPE) cells [6, 22]. To understand VEGF expression, a recombinant adenovirus vector expressing rat VEGF$_{164}$ was constructed and injected into the sub-retinal space. RPE cells increased their expression of VEGF messenger RNA (mRNA), and blood vessels became leaky 10 days post-injection. By 80 days post-injection, new blood vessels had originated from the choriocapillaris, which ultimately led to the formation of choroidal neovascular membranes and the death of photoreceptor cells. This study demonstrated that overexpression of VEGF in the RPE cells can induce vascular leakage, new choroidal blood vessel growth, the development of choroidal neovascularization (CNV), and neural retina degeneration [6]. This is the same process by which AMD has been shown to cause vision loss, suggestive that VEGF overexpression plays a key role in AMD.

In a retrospective study comparing the safety and efficacy of two anti-VEGF agents, bevacizumab and ranibizumab, in the treatment of patients with neovascular AMD, a comparable safety and efficacy in terms of gains in visual acuity and reduction in macular thickness was documented [23, 24] It is likely that a randomized controlled trial, if it can be done, will show that bevacizumab is equivalent to ranibizumab in terms of efficacy and safety [24].

VEGF Inhibition

Currently, there are several approved therapeutic agents (and many more being studied) that employ several unique mechanisms of action to inhibit the VEGF pathway. One approach involves the use of monoclonal antibodies to target either VEGF itself or its receptors. Also, VEGF soluble receptors with high affinity for VEGF have been designed to prevent VEGF from binding to VEGF receptors on ECs. Furthermore, various small molecule tyrosine kinase inhibitors have been developed to inhibit VEGF tyrosine kinase receptors. Two unique classes of drugs are targeting the mRNA used to code for VEGF. One class is designed to target post-transcriptional modification of mRNA and actually prevent the protein translation of VEGF [25].

VEGF Inhibition in the Treatment of AMD

Pegaptanib (Macugen®), an aptamar that binds $VEGF_{165}$ is approved by the FDA for the treatment of wet AMD. Its efficacy and safety analysis were reported in two randomized, sham-controlled clinical trials. These two combined trials are known as the VEGF Inhibition Study in Ocular Neovascularization (VISION), which enrolled 1,186 patients. The patients received either an intraocular injection pegaptanib or a similar sham injection every 6 weeks. Visual acuity (VA) was measured using Snellen eye charts in which patients are asked to identify specific-sized letters or lines at a set distance. Results from the VISION trials indicate that pegaptanib is effective at reducing vision loss compared to sham injection in patients with several types of AMD [26, 27].

Pegaptanib was shown to be a cost-effective treatment for wet AMD in elderly patients as compared to the standard of care in the UK, and as compared to photodynamic therapy (PDT) (verteporfin), and as compared to the standard of care in Canada [28, 29].

Ranibizumab (Lucentis®) was approved for the treatment of wet AMD. Ranibizumab was studied in a 2-year, phase III, double-blind, randomized, sham-controlled study. Patients received either ranibizumab low dose (n=238), ranibizumab high dose (n=240), or a sham injection given intravitreally monthly for

2 years in one eye. The primary outcome of VA was assessed by measuring the number of patients who lost fewer than 15 letters from baseline. The mean VA improved by about seven letters in the ranibizumab group compared with a decline of 10 letters in the sham-injection group ($p < 0.001$). At the study conclusion, 26.1 and 33.3 % of patients in the low and high dose ranibizumab group, respectively, had a VA gain of 15 letters or more, compared with 3.8 % of patients in the sham-injection group ($p < 0.001$) [30, 31].

No trials have been conducted comparing ranibizumab to pegaptanib. However, overall data shows that ranibizumab actually produces an increase in VA from baseline.

Verteporfin PDT was also indicated for wet AMD, and previous to VEGF inhibiting therapy was the treatment of choice in wet AMD. A study compared ranibizumab to verteporfin PDT in a 2-year, multicenter, double-blind, randomized trial where patients received either low or high doses of ranibizumab or verteporfin PDT [32]. Patients receiving ranibizumab had significantly better VA as indicated by more patients losing fewer than 15 letters on Snellen charts, and patients in the ranibizumab group gained 15 or more letters in VA (35.7 % low dose and 40.3 % high dose) compared to the verteporfin group (5.6 %, $p < 0.001$). Severe loss of VA, indicated by a decline of 30 letters or more, occurred among 13.3 % of patients receiving verteporfin compared with none among patients receiving ranibizumab [32].

Ongoing clinical trials are currently investigating other therapies for AMD that target VEGF. A soluble VEGF receptor, VEGF trap, was studied in phase II and III trials as an intravitreal injection. Also, trials examining the systemic and intraocular administration of bevacizumab, a VEGF monoclonal antibody, to treat AMD showed that the maximum tolerated intravenous dose of VEGF trap in this study population was 1.0 mg/kg. This dose resulted in elimination of about 60 % of excess retinal thickness after either single or multiple administrations [33].

Bevasiranib, the first small interfering RNA agent developed for the treatment of neovascular AMD, has demonstrated clinical promise. Bevasiranib targets the production of VEGF protein. It does not affect existing VEGF protein, suggesting that it may offer a synergistic effect when given in combination with anti-VEGF treatments, such as ranibizumab. The safety of bevasiranib has been supported by preclinical and clinical research [34].

Comparison Among Different Pharmacotherapies

The effects of different treatments on serious pigment epithelium detachment in AMD were investigated. Results were significantly better in patients treated with bevacizumab and ranibizumab than in those treated with pegaptanib or with a combination of PDT and intravitreal triamcinolone acetonide. Even with treatment, tears of the RPE or partial flattening of the pigment epithelium detachment always indicated a worse prognosis in eyes with exudative AMD than in eyes with CNV [35].

   Patients with AMD of any lesion type benefit from treatment with either pegaptanib or ranibizumab with regards to VA when compared with sham injection and/or PDT. When comparing pegaptanib and ranibizumab, the evidence was less clear due to the lack of a designed head-to-head comparison [36].

Aflibercept

A pivotal phase III VEGF trap-eye trial in patients with wet AMD showed that aflibercept was non-inferior to ranibizumab in preventing vision loss with comparable vision gains, safety, and perhaps at lower cost than ranibizumab [37]. Aflibercept gained FDA drug approval after two randomized, double-blind phase III trials were conducted: VIEW 1 and VIEW 2 (VEGF trap-eye: Investigation of Efficacy and Safety in Wet Age-related Macular Degeneration [37, 38]). These studies were conducted to measure the safety and efficacy of aflibercept compared to ranizumab, the standard of care for wet AMD [39]. VIEW 1 was the first study conducted, and VIEW 2 followed after strong evidence from VIEW 1 that aflibercept was comparable to the standard of care. Both studies had the same endpoints, treatment group population, and primary outcome measures. The only difference was that VIEW 1 was conducted in North America, whereas VIEW 2 was conducted internationally. Both trials had a set outcome goal at 52 weeks of treatment, and the primary outcome was identified as the percentage of patients who maintained vision at week 52. Maintaining vision was defined as patients who lost less than 15 letters based on the best-corrected visual acuity (BCVA) scale compared to baseline measurements [38]. The safety analysis in both VIEW trials displayed a well-tolerated drug in aflibercept [40]. When compared to ranibizumab, the safety profile was of approximately equivalent measurements [39–41]. The most common serious side effects presented were loss in VA, retinal hemorrhage, and endophtalmitis. These studies provided a foundation for aflibercept approval in the treatment of wet AMD.

Combined Therapies in AMD

The effect of combined PDT and intravitreal injection of bevacizumab in occult CNV with recent disease progression and in CNV due to AMD was investigated [42]. It was concluded that PDT combined with injection of intravitreal bevacizumab was well-tolerated and is effective along with stabilization of VA in 96 % of patients. Further studies are necessary to show the long-term effect of the PDT and anti-VEGF combination therapy [42]. Combination bevacizumab and low dose PDT significantly reduced the number of bevacizumab treatments required over 6 months.

However, this study was powered to examine number of treatments, but not visual acuities, and further studies are required to explore visual outcomes [42, 43].

Overall, PDT has been widely replaced by anti-angiogenesis agents (anti-VEGF) for the first-line therapy for exudative AMD. There is a strong basis for predicting that a combination of PDT and anti-VEGF drugs may address the relative disadvantages of each by improving the response rates and reducing the frequency with which intravitreal injections of anti-VEGF are required. Anti-VEGF drugs may augment the activity of PDT by inhibiting its counterproductive up-regulation of VEGF. Clinical studies of this combination are being advanced in both AMD and in the treatment of certain malignancies.

In a retrospective case series database study (registry), 1,196 patients with CNV due to AMD received one or more combination treatments of bevacizumab (1.25 mg) within 14 days of verteporfin. The use of verteporfin with bevacizumab resulted in vision benefit for most patients [44]. The efficacy and safety of triple therapy consisting of single-session PDT, intravitreal bevacizumab and intravitreal triamcinolone for treatment of neovascular AMD was evaluated in patients with subfoveal CNV secondary to AMD [45]. The study concluded that short-term results of single-session triple therapy suggested that it might be a useful treatment option for neovascular AMD based on its low re-treatment rates, sustainable CNV eradication result, and visual gain achievement. However, the risk and benefits of using intravitreal triamcinolone in addition to combined PDT and intravitreal bevacizumab warrant further evaluation [45].

Tyrosine Kinase Inhibitors

Currently in phase III clinical trials, VEGF trap is a receptor decoy that targets VEGF with higher affinity [32] than ranibizumab and other currently available anti-VEGF agents. Another promising therapeutic strategy is the blockade of VEGF effects by inhibition of the tyrosine kinase cascade downstream from the VEGF receptor; such therapies currently in development include vatalanib, TG100801, pazopanib, AG013958, and AL39324. Small interfering RNA technology-based therapies have been designed to down-regulate the production of VEGF (bevasiranib) or VEGF receptors (AGN211745) by degradation of specific mRNA. Other potential therapies include pigment epithelium-derived factor-based therapies, nicotinic acetylcholine receptor antagonists, integrin antagonists, and sirolimus.

An oral, multi-targeted receptor tyrosine kinase inhibitor, SU11248, that inhibits VEGFR-2, PDGF receptor, and Fms-like tyrosine kinase 3 (FLT3) demonstrated suppression of leakage in an experimental mouse model of CNV caused by AMD [46].

Additionally, inhibition of these tyrosine kinase receptors prevents tumor growth, pathologic angiogenesis, and metastatic progression of cancer [47]. Sunitinib is currently the only FDA-indicated drug for gastrointestinal stromal tumors. Compared with placebo, sunitinib improved time to tumor progression (TTP) and progression-free survival (PFS) time in patients with gastrointestinal stromal tumor who had previously not responded to imatinib [48, 49].

Another tyrosine kinase inhibitor, sorafenib (Nexavar®, BAY 43–9006) [50], also inhibits tumor angiogenesis by blocking the activation of several tyrosine kinase receptors involved in neovascularization and tumor progression, including VEGFR-2, VEGFR-3, PDGFR-β, FLT3, c-Kit and p38-alpha. In addition, sorafenib inhibits the activity of Raf-1 and B-Raf, which are involved in the regulation of endothelial apoptosis [51, 52]. Sorafenib is FDA-approved and has been studied in phase III trials for advanced renal cell carcinoma [51]. In phase III trials, oral sorafenib prolonged PFS compared with placebo in patients with advanced clear-cell renal cell carcinoma in whom first-line therapy had failed. Also, partial responses were significantly higher in the sorafenib group compared to placebo. Treatment was associated with increased adverse events including diarrhea, rash, fatigue, hand-foot skin reactions, hypertension, and cardiac ischemia. In addition, sorafenib significantly increased PFS in patients with advanced renal cell carcinoma in a phase II, placebo-controlled trial [52].

AEE788 is potent, combined inhibitor of both endothelial growth factor (EGF) receptor and VEGF receptor tyrosine kinase family members. *In vitro*, EGF receptor and VEGF receptor phosphorylation was efficiently inhibited, and AEE788 demonstrated anti-proliferative activity against a range of EGF receptor and ErbB2 overexpression cell lines and inhibited the proliferation of EGF- and VEGF-stimulated human umbilical vein ECs [53]. *In vivo*, AEE788 decreased tumor growth in a number of cancer cell lines that overexpress EGFR and/or ErbB2. Oral administration of AEE788 to tumor-bearing mice resulted in high and persistent compound levels in tumor tissue. In addition, AEE788 also inhibited VEGF-induced angiogenesis in a murine implant model [53]. Consequently, AEE788 is currently being studied in phase I clinical trials.

Axitinib (AG013736) is an oral selective inhibitor of VEGF receptors 1, 2, and 3. In a phase II clinical trial, axitinib was studied in patients diagnosed with metastatic renal cell cancer who had failed on previous cytokine-based treatment. The primary endpoint was objective response (based on RECIST criteria), and secondary endpoints were duration of response, TTP, overall survival, safety, pharmacokinetics, and patient-reported health-related quality of life. Results showed 2 out of 52 complete and 21 out of 52 partial responses, with an objective response rate of 44.2 %, and median response duration was 23.0 months [54]. Treatment-related adverse events included diarrhea, hypertension, fatigue, nausea, and hoarseness. Data suggest that axitinib might have clinical benefit in patients with cytokine-refractory metastatic renal cell cancer [54].

Cediranib (AZD2171) is a highly potent ATP-competitive inhibitor of recombinant KDR tyrosine kinase. *In vitro* experiments using human umbilical vein ECs showed AZD2171's ability to inhibit VEGF-stimulated proliferation and KDR phosphorylation and inhibit vessel sprouting in fibroblast and EC models [55]. Additionally, AZD2171 inhibited tumor growth in a mouse xenograft model of colon, lung, prostate, breast, and ovarian cancer [56].

Zactima (ZD6474) is an orally bioavailable inhibitor of VEGFR-2 tyrosine kinase with additional activity against the EGF receptor tyrosine kinase. In preclinical studies, ZD6474 blocked *in vivo* phosphorylation of VEGF and EGF tyrosine kinase receptors and prevented the growth of human cancer cell lines in nude mice xenograft [57]. However, disappointing results from a phase II trial in patients with previously treated metastatic breast cancer were recently made available [58].

Vatalanib (PTK787/ZK 222584) is an orally bioavailable angiogenesis inhibitor targeting all known VEGF receptor tyrosine kinases, including VEGFR-1, VEGFR-2, VEGFR-3, the PDGF receptor tyrosine kinase, and the c-Kit protein tyrosine kinase [59].

Pazopanib (GW786034) is a tyrosine kinase inhibitor that inhibits VEGFR-1, -2, -3, PDGF receptors, and c-Kit. A phase I study demonstrated activity in various types of advanced solid tumors [60]. In a phase II trial, pazopanib resulted in stable disease or partial response in 42 % (25/60) of patients at 12 weeks [61]. Adverse events included hypertension, fatigue, diarrhea, nausea, and proteinuria. Surprisingly, no cases of hand and foot syndrome were reported, and only one case of bleeding occurred. Results appear encouraging, and phase II/III trials are underway in CNV and cancer.

AV-951 (KRN951) is an orally bioavailable tyrosine kinase inhibitor that is specific for VEGF receptors 1, 2 and 3. AV-951 potently inhibited VEGF-induced VEGFR-2 phosphorylation in ECs but not VEGF-independent activation of mitogen-activated protein kinases and proliferation of ECs [62]. Following oral administration to rats, AV-951 decreased the microvessel density within tumor xenograft and decreased VEGFR-2 phosphorylation within tumor endothelium. It also inhibited tumor growth in a wide variety of human tumor xenografts, including lung, breast, colon, ovarian, pancreas, and prostate cancer [62]. In a phase I clinical trial consisting of 40 patients with advanced solid tumors, AV-951 showed promising results. Notably, of the 9 patients with refractory renal cell carcinoma, all achieved either a partial response or stable disease with one patient exhibiting a response lasting more than 30 months [63]. However, a phase II trial in renal cell carcinoma was disappointing, leading to discontinuation of AV-951 [64].

AMG 706 is an orally bioavailable inhibitor of the VEGFR-1, VEGFR-2, VEGFR-3, PDGF receptor, and Kit receptors in preclinical models. AMG 706 inhibited human endothelial cell proliferation induced by VEGF, but not by bFGF *in vitro* [65]. In addition, it inhibited vascular permeability induced by VEGF in mice, and its oral administration inhibited VEGF-induced angiogenesis in the rat corneal model and induced regression of established A431 xenografts [66]. In a phase I trial enrolling 71 patients, the most frequent adverse events were fatigue, diarrhea, nausea, and hypertension [66]. Thirty four patients (61 %) had stable disease (at least through 1 month). In this phase I study of patients with advanced refractory solid tumors, AMG 706 was well-tolerated and there is evidence of antitumor activity [67]. Additional studies of AMG 706 as monotherapy and in combination with various agents are ongoing.

Issues with VEGF Inhibitors

Although VEGF inhibitors represent the culmination of decades of research in the treatment of several disease states, numerous issues still need to be addressed before their true benefit can be known. Measuring the efficacy of VEGF inhibitors is difficult. Although tumor regression has occurred in some cases, angiogenesis inhibitors are not typically cytotoxic; rather they will most likely result in growth stasis. Some of the current criteria we use to define whether a therapy is efficacious may need to be modified. Tumor mass is likely to be a poor indicator of effective therapy with angiogenesis inhibitors.

Monoclonal antibodies have historically been considered the "magic bullet." Consequently, utilizing monoclonal antibodies has become a cornerstone in cytokine-targeting therapies. However, cases have been reported where endogenous antibodies actually target these monoclonal antibodies, rendering them inactive [68, 69]. Therefore, as with any monoclonal antibody, it is likely that these types of reactions will occur with anti-VEGF antibodies. Additionally, pharmacoeconomic analysis for cost-effectiveness is still not well-developed enough to justify these expensive therapies.

In addition, blocking VEGF, or its receptors, may block or potentiate the effects of other ligands. It is likely that these receptors are not specific for VEGF. It is difficult to determine what the long-term effects of blocking VEGF and its receptors may be. In clinical trials, frequent adverse events of most VEGF inhibitors include a dramatic increase in the rate of thromboembolic events [70]. Cancer has been shown to increase the risk of these events alone, without anti-VEGF therapy, and thus it seems that concurrent anticoagulation therapy would be beneficial for many patients. However, other data reveal that bleeding is a common adverse event with these therapies as well. Hopefully, future research can enlighten practitioners as to which patient populations are at risk for an adverse event, and therapy can be tailored accordingly.

Beyond VEGF-targeted Therapies

VEGF inhibitors are a milestone in drug development; in spite of this, the issues discussed above make it unlikely that they will be useful in all patients. VEGF inhibitors do appear to be valuable in many types of cancer, nevertheless, not in all types. Alone or in combination with chemotherapy, VEGF inhibitors in trials have had mixed results. For this reason, it would be helpful to have diagnostic testing available to determine which patients would benefit from therapy. Perhaps certain patient populations will be identified that would benefit most by targeting a specific angiogenic growth factor or a specific drug class targeting that growth factor. Moreover, it seems necessary to identify potential antagonism/synergy between certain agents, thus allowing us to predict the most efficacious combinations and enabling practitioners to overcome redundancies that are built into the angiogenesis process. Other novel therapies with different targets may potentially have fewer adverse events and benefit certain populations that cannot receive anti-VEGF therapy.

Conclusions

AMD treatment before the year 2000 was limited to focal laser photocoagulation in an effort to limit the spread of CNV, although it is a destructive procedure and produces a permanent scar. It turned out to be only really viable for treating extrafoveal CNV, and even then it was not entirely effective. In the year 2000, PDT with verteporfin represented the first treatment proven to reduce the risk of vision loss in sub-foveal CNV. However, its efficacy was limited to classic or small CNV, and even though it is a relatively nondestructive form of therapy, it failed to improve vision in patients with AMD in clinical trials.

VEGF was shown to play an important role in promoting angiogenesis and vascular leakage, CNV infiltration, and fluid accumulation in neovascular AMD. Therefore, inhibiting VEGF held the promise of more effectively controlling neovascular AMD. Extensive pre-clinical and clinical research led to the FDA approval of pegaptanib in 2004 and ranibizumab in 2006. Off-label usage of bevacizumab has also become fairly standard. The VA gains recorded with ranibizumab proved particularly exciting, and ranibizumab has become the gold standard for AMD therapy. However, as with many new therapies, there are unresolved issues including safety, cost, and dosing frequency. Further preclinical and clinical investigations are in progress in the inhibition of VEGF-mediated effects at various levels and beyond VEGF.

References

1. Mousa SA (2000) Mechanisms of angiogenesis in vascular disorders: potential therapeutic targets. In: Mousa SA (ed) Angiogenesis inhibitors & stimulators: potential therapeutic implications. Landes Bioscience (Autsin, TX), pp 1–12, Chapter 1
2. Relf M, LeJeune S, Scott P et al (1997) Expression of the angiogenic factors vascular endothelial growth factor, acidic and basic fibroblast growth factor, tumor growth factor-β-1, platelet-derived endothelial cell growth factor, placenta growth factor, and pleiotrophin in human primary breast cancer and its relation to angiogenesis. Cancer Res 57:963–969
3. Ferrara N, Alitalo K (1999) Clinical applications of angiogenic growth factors and their inhibitors. Nat Med 5:1359–1364
4. Mousa SA, Mousa AS (2004) Angiogenesis inhibitors: current & future directions. Curr Pharm Des 10(1):1–9
5. De Jong PT (2006) Age-related macular degeneration. N Engl J Med 355(14):1474–1485
6. Spilsbury K, Garrett KL, Shen WY, Constable IJ, Rakoczy PE (2000) Overexpression of vascular endothelial growth factor (VEGF) in the retinal pigment epithelium leads to the development of choroidal neovascularization. Am J Pathol 157(1):135–144

7. Fink AM, Cauza E, Hassfeld W, Dunky A, Bayer PM, Jurecka W, Steiner A (2007) Vascular endothelial growth factor in patients with psoriatic arthritis. Clin Exp Rheumatol 25(2):305–308
8. Murakami M, Iwai S, Hiratsuka S, Yamauchi M, Nakamura K, Iwakura Y, Shibuya M (2006) Signaling of vascular endothelial growth factor receptor-1 tyrosine kinase promotes rheumatoid arthritis through activation of monocytes/macrophages. Blood 108(6):1849–1856
9. Kerbel R, Folkman J (2002) Clinical translation of angiogenesis inhibitors. Nat Rev Cancer 2:727–739
10. Marx J (2002) Cancer research: obstacle for promising cancer therapy. Science 295:1444
11. Casanovas O, Hicklin D, Bergers G, Hanahan D (2005) Drug resistance by evasion of antiangiogenic targeting of VEGF signaling in late-stage pancreatic islet tumors. Cancer Cell 4(8):299–309
12. Nemunaitis J, Dillman RO, Schwarzenberger PO et al (2006) Phase II study of belagenpumatucel-L, a transforming growth factor beta-2 antisense gene-modified allogeneic tumor cell vaccine in non–small-cell lung cancer. J Clin Oncol 24(29):4721–4730
13. Kelly RJ, Giaccone G (2011) Lung cancer vaccines. Cancer J 17(5):302–308
14. Suzuki E, Kapoor V, Cheung H, Ling LE, DeLong PA, Kaiser LR, Albelda SM (2004) Soluble type II transforming growth factor-ß receptor inhibits established murine malignant mesothelioma tumor growth by augmenting host antitumor immunity. Clin Cancer Res 10(17):5907–5918
15. Brown LF, Detmar M, Claffey K, Nagy JA, Feng D, Dvorak AM, Dvorak HF (1997) Vascular permeability factor/vascular endothelial growth factor: a multifunctional angiogenic cytokine. EXS 79:233–269
16. Joukov V, Kaipainen A, Jeltsch M et al (1997) Vascular endothelial growth factors VEGF-B and VEGF-C. J Cell Physiol 173(2):211–215
17. Tischer E, Mitchell R, Hartman T, Silva M, Gospodarowicz D, Fiddes JC, Abraham JA (1991) The human gene for vascular endothelial growth factor: multiple protein forms are encoded through alternative exon splicing. J Biol Chem 266(18):11947–11954
18. Ferrara N (1999) Molecular and biological properties of vascular endothelial growth factor. J Mol Med 77(7):527–543
19. Dvorak HF, Nagy JA, Feng D et al (1999) Vascular permeability factor/vascular endothelial growth factor and the significance of microvascular hyperpermeability in angiogenesis. Curr Top Microbiol Immunol 237:97–132
20. Senger DR, Van De Water L, Brown LF et al (1993) Vascular permeability factor (VPF, VEGF) in tumor biology. Cancer Metastasis Rev 12(3–4):303–324
21. Pettersson A, Nagy JA, Brown LF et al (2000) Heterogeneity of the angiogenic response induced in different normal adult tissues by vascular permeability factor/vascular endothelial growth factor. Lab Invest 80(1):99–115
22. Ford KM, D'Amore PA (2012) Molecular regulation of vascular endothelial growth factor expression in the retinal pigment epithelium. Mol Vis 18:519–527
23. Landa G, Amde W, Doshi V, Ali A, McGevna L, Gentile RC, Muldoon TO, Walsh JB, Rosen RB (2009) Comparative study of intravitreal bevacizumab (avastin) versus ranibizumab (lucentis) in the treatment of neovascular age-related macular degeneration. Ophthalmologica 223(6):370–375
24. Schouten JS, La Heij EC, Webers CA, Lundqvist IJ, Hendrikse F (2009) A systematic review on the effect of bevacizumab in exudative age-related macular degeneration. Graefes Arch Clin Exp Ophthalmol 247(1):1–11
25. Hirawat S, Elfring GL, Northcutt VJ, Paquette N (2007) Phase I studies assessing the safety, PK, and VEGF-modulating effects of PTC299, a novel VEGF. J Clin Oncol 25(18S):3562
26. Gragoudas ES, Adamis AP, Cunningham ET Jr, Feinsod M, Guyer DR (2004) VEGF inhibition study in ocular neovascularization clinical trial group. Pegaptanib for neovascular age-related macular degeneration. N Engl J Med 351(27):2805–2816
27. Chakravarthy U, Adamis AP, VEGF Inhibition Study in Ocular Neovascularization (V.I.S.I.O.N.) Clinical Trial Group et al (2006) Year 2 efficacy results of pegaptanib for neovascular age-related macular degeneration. Ophthalmology 113(9):1508–1525

28. Wolowacz SE, Roskell N, Kelly S, Maciver FM, Brand CS (2007) Cost effectiveness of pegaptanib for the treatment of age-related macular degeneration in the UK. Pharmacoeconomics 25(10):863–879
29. Earnshaw SR, Moride Y, Rochon S (2007) Cost-effectiveness of pegaptanib compared to photodynamic therapy with verteporfin and to standard care in the treatment of subfoveal wet age-related macular degeneration in Canada. Clin Ther 29(9):2096–2106
30. Rosenfeld PJ, Brown DM, Heier JS et al (2006) Ranibizumab for neovascular age-related macular degeneration. N Engl J Med 355(14):1419–1431
31. Heier JS, Antoszyk AN, Pavan P et al (2006) Ranibizumab for treatment of neovascular age-related macular degeneration – a phase I/II multicenter, controlled, multidose study. Ophthalmology 113(N4):633–642
32. Brown DM, Kaiser PK, Michels M et al (2006) Ranibizumab versus verteporfin for neovascular age-related macular degeneration. N Engl J Med 355(14):1432–1444
33. Nguyen QD, Shah SM, Hafiz G, Quinlan E, Sung J, Chu K, Cedarbaum JM, Campochiaro PA, CLEAR-AMD 1 Study Group (2006) A phase I trial of an IV-administered vascular endothelial growth factor trap for treatment in patients with choroidal neovascularization due to age-related macular degeneration. Ophthalmology 113(9):1522.e1–1522.e14
34. Singerman L (2009) Combination therapy using the small interfering RNA bevasiranib. Retina 29(6 Suppl):S49–S50
35. Lommatzsch A, Heimes B, Gutfleisch M, Spital G, Zeimer M, Pauleikhoff D (2009) Serious pigment epithelial detachment in age-related macular degeneration: comparison of different treatments. Eye 23(12):2163–2168
36. Maier MM, Feucht N, Fiore B, Winkler von Mohrenfels C, Kook P, Fegert C, Lohmann C (2009) Photodynamic therapy with verteporfin combined with intravitreal injection of ranibizumab for occult and classic CNV in AMD. Klin Monatsbl Augenheilkd 226(6):496–502
37. Vascular Endothelial Growth Factor (VEGF) (2012) Trap-eye: investigation of efficacy and safety in wet Age-Related Macular Degeneration (AMD) (VIEW 2). http://clinicaltrials.gov/show/NCT00637377. Accessed 13 Mar 2013
38. Vascular Endothelial Growth Factor (2012) VEGF trap-eye: investigation of efficacy and safety in wet Age-Related Macular Degeneration (AMD) (VIEW1). http://clinicaltrials.gov/show/NCT00509795. Accessed 13 Mar 2013
39. Nguyen QD et al (2012) Evaluation of very high- and very low-dose intravitreal aflibercept in patients with neovascular age-related macular degeneration. J Ocul Pharmacol Ther 28:581–588
40. Frampton JE (2012) Aflibercept for intravitreal injection: in neovascular age-related macular degeneration. Drugs Aging 29(10):839–846
41. Xu D, Kaiser PK (2013) Intravitreal aflibercept for neovascular age-related macular degeneration. Immunotherapy 5(2):121–130
42. Potter MJ, Claudio CC, Szabo SM (2010) A randomised trial of bevacizumab and reduced light dose photodynamic therapy in age-related macular degeneration: the VIA study. Br J Ophthalmol 94:174–179
43. Busch T (2009) Approaches toward combining photodynamic therapy with pharmaceuticals that alter vascular microenvironment. Retina 29(6 Suppl):S36–S38
44. Kaiser PK, Registry of Visudyne AMD Therapy Writing Committee, Boyer DS, Garcia R, Hao Y, Hughes MS, Jabbour NM, Kaiser PK, Mieler W, Slakter JS, Samuel M, Tolentino MJ, Roth D, Sheidow T, Strong HA (2009) Verteporfin photodynamic therapy combined with intravitreal bevacizumab for neovascular age-related macular degeneration. Ophthalmology 116(4):747–755
45. Yip PP, Woo CF, Tang HH, Ho CK (2009) Triple therapy for neovascular age-related macular degeneration using single-session photodynamic therapy combined with intravitreal bevacizumab and triamcinolone. Br J Ophthalmol 93(6):754–758
46. Takahashi H, Obata R, Tamaki Y (2006) A novel vascular endothelial growth factor receptor 2 inhibitor, SU11248, suppresses choroidal neovascularization in vivo. J Ocul Pharmacol Ther 22(4):213–218

47. Faivre S, Delbaldo C, Vera K et al (2006) Safety, pharmacokinetic, and antitumor activity of SU11248, a novel oral multitarget tyrosine kinase inhibitor, in patients with cancer. J Clin Oncol 24(1):25–35
48. Demetri GD, VanOosterom AT, Garrett CR et al (2006) Efficacy and safety of sunitinib in patients with advanced gastrointestinal stromal tumor after failure of imatinib: a randomised controlled trial. Lancet 368:1329–1338
49. Strumberg D, Richly H, Hilger RA et al (2005) Phase I clinical and pharmacokinetic study of the novel Raf kinase and vascular endothelial growth factor receptor inhibitor BAY 43–9006 in patients with advanced refractory solid tumors. J Clin Oncol 23(5):965–972
50. Awada A, Hendlisz A, Gil T et al (2005) Phase I safety and pharmacokinetics of BAY 43–9006 administered for 21 days on/7 days off in patients with advanced, refractory solid tumors. Br J Cancer 92(10):1855–1861
51. Escudier B, Eisen T, Stadler WM et al (2007) Sorafenib in advanced clear-cell renal-cell carcinoma. N Engl J Med 356(2):125–134
52. Ratain MJ, Eisen T, Stadler WM et al (2005) Final findings from a phase II, placebo-controlled, randomized discontinuation trial (RDT) of sorafenib (BAY 43–9006) in patients with advanced renal cell carcinoma (RCC) (4544). J Clin Oncol 23(16 suppl):388s
53. Traxler P, Allegrini PR, Brandt R et al (2004) AEE788: a dual family epidermal growth factor receptor/ErbB2 and vascular endothelial growth factor receptor tyrosine kinase inhibitor with antitumor and antiangiogenic activity. Cancer Res 64:4931–4941
54. Rixe O, Bukowski RM, Michaelson MD et al (2007) Axitinib treatment in patients with cytokine-refractory metastatic renal-cell cancer: a phase II study. Lancet Oncol 8(11):975–984
55. Wedge SR, Kendrew J, Hennequin LF, Valentine PJ (2005) AZD2171: a highly potent, orally bioavailable, vascular endothelial growth factor receptor-2 tyrosine kinase inhibitor for the treatment of cancer. Cancer Res 65(10):4389–4395
56. Drevs J, Siegert P, Medinger M et al (2007) Phase I clinical study of AZD2171, an oral vascular endothelial growth factor signaling inhibitor, in patients with advanced solid tumors. J Clin Oncol 25(21):3045–3054
57. Ciardiello F, Caputo R, Damiano V et al (2003) Antitumor effects of ZD6474, a small molecule vascular endothelial growth factor receptor tyrosine kinase inhibitor, with additional activity against epidermal growth factor receptor tyrosine kinase. Clin Cancer Res 9:1546–1556
58. Miller KD, Trigo J, Wheeler C et al (2005) A multicenter phase II trial of ZD6474, a vascular endothelial growth factor receptor-2 and epidermal growth factor receptor tyrosine kinase inhibitor, in patients with previously treated metastatic breast cancer. Clin Cancer Res 11(9):3369–3376
59. Thomas AL, Trarbach T, Bartel C et al (2007) A phase IB, open-label dose-escalating study of the oral angiogenesis inhibitor PTK787/ZK 222584 (PTK/ZK), in combination with FOLFOX4 chemotherapy in patients with advanced colorectal cancer. Ann Oncol 18(4):782–788
60. Hutson TE, Davis ID, Machiels JP et al (2007) Pazopanib (GW786034) is active in metastatic renal cell carcinoma (RCC): interim results of a phase II randomized discontinuation trial (RDT). J Clin Oncol 25(18S):5031
61. Suttle AB, Hurwitz H, Dowlati A et al (2004) Pharmacokinetics (PK) and tolerability of GW786034, a VEGFR tyrosine kinase inhibitor, after daily oral administration to patients with solid tumors. J Clin Oncol 22:3054
62. Nakamura K, Taguchi E, Miura T et al (2006) KRN951, a highly potent inhibitor of vascular endothelial growth factor receptor tyrosine kinases, has antitumor activities and affects functional vascular properties. Cancer Res 66(18):9134–9142
63. Eskens FA, Planting A, Van Doorn L et al (2006) An open-label phase I dose escalation study of KRN951, a tyrosine kinase inhibitor of vascular endothelial growth factor receptor 2 and 1 in a 4 week on, 2 week off schedule in patients with advanced solid tumors. J Clin Oncol [supplement ASCO Annual Meeting] 24(18S):2034
64. Nosov DA, Esteves B, Lipatov ON, Lyulko AA, Anischenko AA, Chacko RT, Doval DC, Strahs A, Slichenmyer WJ, Bhargava P (2012) Antitumor activity and safety of tivozanib (AV-951) in a phase II randomized discontinuation trial in patients with renal cell carcinoma. J Clin Oncol 30(14):1678–1685

65. Rosen LS, Kurzrock R, Mulay M et al (2007) Safety, pharmacokinetics, and efficacy of AMG 706, an oral multikinase inhibitor, in patients with advanced solid tumors. J Clin Oncol 25(17):2369–2376
66. Polverino A, Coxon A, Starnes C et al (2006) AMG 706, an oral, multikinase inhibitor that selectively targets vascular endothelial growth factor, platelet-derived growth factor, and kit receptors, potently inhibits angiogenesis and induces regression in tumor xenografts. Cancer Res 66(17):8715–8721
67. Rosen R, Kurzrock E, Jackson L et al (2005) Safety and pharmacokinetics of AMG 706 in patients with advanced solid tumors. J Clin Oncol 23(16S):3013
68. Carter P (2001) Improving the efficacy of antibody-based cancer therapies. Nat Rev Cancer 1:118–129
69. Vedula SS, Krzystolik MG (2008) Antiangiogenic therapy with anti-vascular endothelial growth factor modalities for neovascular age-related macular degeneration. Cochrane Database Syst Rev Apr 16(2):CD005139
70. Costagliola C, Agnifili L, Arcidiacono B, Duse S, Fasanella V, Mastropasqua R, Verolino M, Semeraro F (2012) Systemic thromboembolic adverse events in patients treated with intravitreal anti-VEGF drugs for neovascular age-related macular degeneration. Expert Opin Biol Ther 12(10):1299–1313

Chapter 14
Application of Nanotechnology to Prevent Tumor Angiogenesis for Therapeutic Benefit

Dhruba J. Bharali and Shaker A. Mousa

Abstract Despite few breakthroughs in cancer treatment over the past several years, cancer remains one of the leading causes of death worldwide. Most of the recent modalities in cancer research heavily depend on invasive procedures (i.e. random biopsies, surgery) and crude, non specific techniques such as irradiation and the use of chemotherapeutic agents. Therefore, it is essential to look for an alternative technique that can effectively target tumor angiogenesis and that has potential clinical relevance. Targeting the tumor angiogenesis to treat cancer is an intense area of research that the medical community has been engaging in for several decades, based on the well-established fact that a tumor cannot grow beyond 1–2 mm in size without angiogenesis. A nanomedicine approach of targeting various angiogenic factors to impair tumor angiogenesis might be an alternative to conventional therapy. In this chapter we discuss the use of nanotechnology to carry different anti-angiogenic agents and therapeutic genes to target tumor vasculature for tumor regression.

Cancer is a disease that affects millions of Americans and touches almost every American family. Despite tremendous efforts by researchers, cancer remains one of the topmost causes of death in the United States and globally. According to the World Health Organization, cancer accounted for upwards of 7.6 million deaths in 2008 worldwide and is the second leading global killer, accounting for 13 % of all deaths [1]. Deaths from cancer in the world are projected to continue rising, with an estimated 13.1 million people dying in 2030 [1, 2]. According to the American Cancer Society, an estimated 1,638,910 new cancer cases are expected to be diagnosed and around 577,190 people will die from cancer in the US in 2012 (which is more than 1,500 per day) [2]. Although there have been a few advances in cancer treatment over the past several decades, current diagnostic and therapeutic approaches

D.J. Bharali • S.A. Mousa (✉)
The Pharmaceutical Research Institute at Albany College of Pharmacy
and Health Sciences, Rensselaer, NY, USA
e-mail: shaker.mousa@acphs.edu

S.A. Mousa and P.J. Davis (eds.), *Angiogenesis Modulations in Health and Disease:*   173
Practical Applications of Pro- and Anti-angiogenesis Targets,
DOI 10.1007/978-94-007-6467-5_14, © Springer Science+Business Media Dordrecht 2013

rely predominantly on invasive procedures (i.e. random biopsies, surgery) and crude, non specific techniques such as irradiation and chemotherapeutic agents. Thus, cancer continues to be almost uniformly fatal, and current therapeutic modalities have yet to significantly improve the dismal prognosis of this disease. Therefore, it is essential to look for an alternative technology like nanotechnology, which has potential to address many of the challenges in cancer treatment. Over the past few years, evidence from the scientific and medical research has demonstrated that nanotechnology and nanomedicine have tremendous potential to profoundly impact numerous aspects of cancer diagnosis and treatment.

Angiogenesis, the physiological process of the development and propagation of new blood vessels, is one of most crucial processes that is needed for the survival and mass development in tumors. It is well-established that a tumor cannot grow beyond 1–2 mm in size without angiogenesis. The main advantages of targeting angiogenesis to treat cancer include better drug accessibility to the tumor vascularization rather than to the solid tumor mass, less probability of being drug-resistant, higher efficacy, wider applicability to various cancer types, and fewer cytotoxic effects to normal cells. Up-regulation of various growth factors like vascular endothelial growth factor (VEGF), angioproteins, and integrins are considered to be the major cause of triggering and nurturing the tumor angiogenesis. Though there have been many therapeutic strategies aiming to stop the progress of the tumor vascularization, most of them are preclinical studies, and few of them are a clinical reality. Therefore, it is essential to look for an alternative, ultramodern technique that can effectively target tumor angiogenesis and that has a potential clinical relevance. A nanomedicine approach of targeting various angiogenic factors might be an alternative to conventional therapy to impair tumor angiogenesis.

There are many integrins that play a vital role in tumor initiation, propagation, and metastasis. $\alpha v \beta 3$ and $\alpha v \beta 5$ were the first integrins that were targeted to inhibit tumor angeogenesis, and since then researchers have been targeting a wide array of integrins including $\alpha 1 \beta 1$, $\alpha 2 \beta 1$, $\alpha v \beta 3$, and $\alpha 5 \beta 1$ using different techniques to suppress tumor angiogenesis. In early studies it was reported that an $\alpha v \beta 3$-targeted nanoparticles-based imaging probe can readily accumulate in tumor vascularization, giving them the capacity to be imaged [3, 4]. In an early study by Arap et al., doxorubicin bound to an αv integrin-binding arg-gly-asp (RGD) motif to increase efficacy of the drug was tested in breast cancer xenografts in nude mice [5]. Since then numerous strategies including several nanoparticle-mediated $\alpha v \beta 3$-targeting nanoparticles containing different therapeutic agents have been tested. Murphy and co-workers synthesized $\alpha v \beta 3$-targeting nanoparticles with a capacity of delivering doxorubicin to the tumor vascularizations [6]. These nanoparticles with RGD and containing distearoylphosphatidylcholine (DSPC), cholesterol dioleoylphosphatidylethanolamine (DOPE), and distearoylphosphatidylethanolamine (DSPE)-mPEG2000 were able to incorporate doxorubicin successfully. They controlled metastasis of pancreatic and renal cell carcinoma in an orthotopic model. The RGD-guided nanoparticles carrying doxorubicin to the tumor vascularization were able to increase the dose response of the drug up to 15-fold when compared to free doxorubicin [6]. Thus, it can be anticipated that the targeted delivery of a cytotoxic drug like doxrobucin not only has the potential to decrease adverse side effects by decreasing the dose limit but also at the same time has the capability to restrict tumor metastasis with a decreased dose of drug.

Zhang et al. synthesized RGD-modified PEGylated polyamidoamine (PAMAM) dendrimer conjugated with doxorubicin and evaluated it in an orthotopic murine model of C6 glioma [7]. They found that when administered intravenously, nano-particle conjugates had a significantly prolonged half-life as well as exhibited a higher accumulation in brain tumor compared to normal brain tissue. They used two different kinds of conjugation techniques to conjugate the dendrimer named as RGD-PPSD and RGD-PPCD where RGD-PPCD shows a 2-fold lower accumula-tion in the tumor than RGD-PPSD. However, it was found that the animals treated with RGD-PPCD conjugated to doxorubicin had a much higher survival rate than RGD-PPCD. Apart from these studies, there are continuous efforts by others in nanoparticles-mediated therapy/imaging research for tumor angiogenesis either to treat or image various cancers in different animal models by targeting tumor neovsacularization. A list of selected studies is in Table 14.1.

Recent studies from our laboratory have shown that tetraiodothyroacetic acid (tetrac), a deaminated thyroid hormone analogue, is capable of preventing the binding of T4 to $\alpha v\beta 3$ at the cell membrane and exerted profound inhibitory effects on cellular proliferation and angiogenesis [18–21]. We have demonstrated the feasibility of using tetrac as an anti-cancer drug *in vivo* and *in vitro* [18–21]. Our laboratory has designed and synthesized a novel nanoformulation made up of PLGA nanopar-ticles with tetrac covalently linked to their surfaces. These nanoparticles target tetrac to the plasma membrane via interaction with the integrin $\alpha v\beta 3$ receptor, providing us with a useful tool and therapeutic means to investigate the anti-angiogenic and anti-proliferative actions of tetrac initiated at the plasma membrane integrin receptor. Tetrac immobilized by covalent bonding on the surface of the nanoparticles might serve as a ligand for the recognition of the integrin $\alpha v\beta 3$ expressed on cancer cells, or its neovascularization site could antagonize plasma membrane receptor-mediated tumor proliferation and angiogenesis processes.

The major advantage of tetrac-nanoparticle formulations is that a nanoparticulate carrier system may restrict entry of tetrac to the cell nucleus, and thus the unwanted genomic effect of this molecule can be avoided. In a series of publications we have shown either equivalent or superior anti-cancer efficacy of tetrac-conjugated nanoparticles in different *in vivo* tumor models including renal cell carcinoma [22], medullary cell carcinoma [23], follicular cell carcinoma [24], and non-small lung cancer cell [25]. In a recent study we also showed that this nanoformulation is highly effective for reversing the development of drug resistance in a doxorubicin-resistant breast cancer tumor model [26]. We strongly believe that it is crucial to study novel, alternative nanoparticles-based technology to treat chemoresistant breast cancer. The development of resistance to chemotherapy represents an adaptive biological response by tumor cells that leads to treatment failure and patient relapse. Unless this problem is solved, cancer will be able to develop resistance to virtually any drug.

D.J. Bharali and S.A. Mousa

Table 14.1 Selected studies showing different nanoparticles-mediated targeted tumor angiogenesis therapeutics/diagnostics in preclinical settings

Type of nanoparticles	Targeting moiety	Therapeutics	Imaging	Therapy	Animal model	Reference
Lipid-based, containing gadolinium	αvβ3-integrin peptido-mimetic antagonist	None	MRI	No	Rabbits implanted with Vx-2 tumors	[4]
Hollow gold nanospheres (HAuNS)	Cyclic RGD peptide c(KRGDf)	Photothermal ablation (PTA)	PET	Yes	Both glioma and angiogenic blood vessels	[8]
Polyethylene glycol–polylactic acid (PEG-PLA)	None	TNP-470	None	Yes	Mice (corneal micropocket angiogenesis and intrasplenic model for induction of liver metastasis)	[9]
Poly-(lactic-co-glycolic acid) (PLGA)	None	Combretastatin-A4 and Doxorubicin	None	Yes	Mice (B16/F10 melanomas or Lewis lung carcinoma)	[10]
Integrin-targeted (ITNP)	Integrin antagonist (BisT-PE-EDTA-IA)	None	Fluorescence microscopy	No	SCC 7, murine squamous cell carcinoma	[11]
Hexadentate-polyD,L-lactic acid-co-glycolic		PD98059, Cisplatin	Fluorescence microscopy	Yes	Melanoma-bearing mice	[12]

Quantum dots (QDs)	RGD peptides	None	None	No	U87MG tumor-bearing mice	[13]
Gold	Cyclic RGD peptides	None	Micro-SPECT/CT	No	Athymic mice with C6-induced tumors	[14]
PLGA	RGD peptide	Paclitaxel and combretastatin	None	Yes	Mice	[15]
Nanographene oxide (GO)	TRC105, a monoclonal antibody	None	PET	No	4T1 murine breast tumor-bearing mice	[16]
Iron oxide nanoworms	Tumor-penetrating peptide iRGD	α-helical amphipathic peptide $_D$[KLAKLAK]$_2$	MRI	Yes	Mice bearing orthotopic glioblastoma tumors	[17]

Abbreviations: *PET* positron emission tomography, *MRI* magnetic resonance imaging, *NIRF* near-infrared fluorescence

The concept of tumor regression and growth inhibition by delivering an appropriate therapeutic gene is another area of intense research. A nanoparticles-mediated gene therapy system has an advantage compared to its counterpart viral vector because of its less cumbersome synthesis method, less immunogenicity, and relatively safer history [27, 28]. In a study by Hood et al. [29], the possibility of targeted gene delivery to the angiogenic tumor vasculature in a mouse model was shown. The nanoparticles synthesized were lipid-based nanoparticles obtained by a combination of self assembly and polymerization, and they were not only capable of carrying a mutant gene Rag gene (*ATPμ-Raf-1*), but also capable of incorporating an $\alpha v \beta 3$ targeting moiety like the LM-609 antibody. It was observed that this nanoparticle formulation was capable of successfully delivering *ATPμ-Raf-1* to the tumor vasculature and which in return was capable of interfering with the signaling cascades of two important angiogenic growth factors, bFGF and VEGF [29]. In another study by Kim et al. [30], a PEGylated polyethyleneimine nanocarrier system comprising PEI-g-PEG-RGD and incorporating a therapeutic gene encoding sFLT1 was designed and synthesized. This nanocomplex containing sFLT1 (a potent VEGF antagonist) efficiently inhibits the proliferation of endothelial cells by obstructing the binding of VEGF to the FLT1 receptor. In recent years there have been more examples of nanoparticles-mediated therapeutic gene delivery [31, 32] and small interfering RNA (siRNA)/short hairpin RNA (shRNA) delivery to target tumor vasculature [33–38]. These genetic materials in return inhibit tumor angiogenesis either by inducing therapeutic genes or silencing unwanted genes responsible for tumor angiogensis.

In summary, nanoparticles-mediated treatment of cancer by targeting tumor vasculatures is one of the most attractive approaches to cancer treatment because of its wide applicability to various cancers, easy drug access to the vasculature, and lower drug-resistance probability compared to conventional cancer therapies. The advantage of the nanoparticles system is that it has not only the capacity to carry a therapeutic payload, but it is also capable of carrying different imaging probes for site-specific delivery. Thus, the versatility of the multifaceted nanocarrier system allows researchers to use an additional imaging approach to visualize and understand the mechanism of action of the drug/nanoparticles in real time. This may allow the precise and targeted delivery to various tumor sites including tumor vasculatures. In recent years, rapid approval of many of the nanoparticles-mediated therapies and anti-angiogenic agents by the US Food and Drug Administration has raised hopes for this technology to bring about a paradigm change in cancer treatment in the clinic.

References

1. World Health Organization (2013) Cancer fact sheet no. 297. http://www.who.int/mediacentre/factsheets/fs297/en/index.html. Accessed 11 Mar 2013
2. American Cancer Society (2012) Cancer facts and figures 2012. http://www.cancer.org/research/cancerfactsfigures/cancerfactsfigures/cancer-facts-figures-2012. Accessed 11 Mar 2013

3. Sipkins DA, Cheresh DA, Kazemi MR, Nevin LM, Bednarski MD, Li KC (1998) Detection of tumor angiogenesis in vivo by alphaVbeta3-targeted magnetic resonance imaging. Nat Med 4:623–626

4. Winter PM, Caruthers SD, Kassner A, Harris TD, Chinen LK, Allen JS, Lacy EK, Zhang H, Robertson JD, Wickline SA, Lanza GM (2003) Molecular imaging of angiogenesis in nascent Vx-2 rabbit tumors using a novel alpha(nu)beta3-targeted nanoparticle and 1.5 tesla magnetic resonance imaging. Cancer Res 63:5838–5843

5. Arap W, Pasqualini R, Ruoslahti E (1998) Cancer treatment by targeted drug delivery to tumor vasculature in a mouse model. Science 279:377–380

6. Murphy EA, Majeti BK, Barnes LA, Makale M, Weis SM, Lutu-Fuga K, Wrasidlo W, Cheresh DA (2008) Nanoparticle-mediated drug delivery to tumor vasculature suppresses metastasis. Proc Natl Acad Sci U S A 105:9343–9348

7. Zhang L, Zhu S, Qian L, Pei Y, Qiu Y, Jiang Y (2011) RGD-modified PEG-PAMAM-DOX conjugates: in vitro and in vivo studies for glioma. Eur J Pharm Biopharm 79:232–240

8. Lu W, Melancon MP, Xiong C, Huang Q, Elliott A, Song S, Zhang R, Flores LG 2nd, Gelovani JG, Wang LV, Ku G, Stafford RJ, Li C (2011) Effects of photoacoustic imaging and photothermal ablation therapy mediated by targeted hollow gold nanospheres in an orthotopic mouse xenograft model of glioma. Cancer Res 71:6116–6121

9. Benny O, Fainaru O, Adini A, Cassiola F, Bazinet L, Adini I, Pravda E, Nahmias Y, Koirala S, Corfas G, D'Amato RJ, Folkman J (2008) An orally delivered small-molecule formulation with antiangiogenic and anticancer activity. Nat Biotechnol 26:799–807

10. Sengupta S, Eavarone D, Capila I, Zhao G, Watson N, Kiziltepe T, Sasisekharan R (2005) Temporal targeting of tumour cells and neovasculature with a nanoscale delivery system. Nature 436:568–572

11. Xie J, Shen Z, Li KC, Danthi N (2007) Tumor angiogenic endothelial cell targeting by a novel integrin-targeted nanoparticle. Int J Nanomedicine 2:479–485

12. Basu S, Harfouche R, Soni S, Chimote G, Mashelkar RA, Sengupta S (2009) Nanoparticle-mediated targeting of MAPK signaling predisposes tumor to chemotherapy. Proc Natl Acad Sci U S A 106:7957–7961

13. Cai W, Chen K, Li ZB, Gambhir SS, Chen X (2007) Dual-function probe for PET and near-infrared fluorescence imaging of tumor vasculature. J Nucl Med 48:1862–1870

14. Morales-Avila E, Ferro-Flores G, Ocampo-Garcia BE, De Leon-Rodriguez LM, Santos-Cuevas CL, Garcia-Becerra R, Medina LA, Gomez-Olivan L (2011) Multimeric system of 99mTc-labeled gold nanoparticles conjugated to c[RGDfK(C)] for molecular imaging of tumor alpha(v)beta(3) expression. Bioconjug Chem 22:913–922

15. Wang Z, Chui WK, Ho PC (2011) Nanoparticulate delivery system targeted to tumor neovasculature for combined anticancer and antiangiogenesis therapy. Pharm Res 28:585–596

16. Hong H, Yang K, Zhang Y, Engle JW, Feng L, Yang Y, Nayak TR, Goel S, Bean J, Theuer CP, Barnhart TE, Liu Z, Cai W (2012) In vivo targeting and imaging of tumor vasculature with radiolabeled, antibody-conjugated nanographene. ACS Nano 6:2361–2370

17. Agemy L, Friedmann-Morvinski D, Kotamraju VR, Roth L, Sugahara KN, Girard OM, Mattrey RF, Verma IM, Ruoslahti E (2011) Targeted nanoparticle enhanced proapoptotic peptide as potential therapy for glioblastoma. Proc Natl Acad Sci U S A 108:17450–17455

18. Glinskii AB, Glinsky GV, Lin HY, Tang HY, Sun M, Davis FB, Luidens MK, Mousa SA, Hercbergs AH, Davis PJ (2009) Modification of survival pathway gene expression in human breast cancer cells by tetraiodothyroacetic acid (tetrac). Cell Cycle 8:3554–3562

19. Rebbaa A, Chu F, Davis FB, Davis PJ, Mousa SA (2008) Novel function of the thyroid hormone analog tetraiodothyroacetic acid: a cancer chemosensitizing and anti-cancer agent. Angiogenesis 11:269–276

20. Mousa SA, Bergh JJ, Dier E, Rebbaa A, O'Connor LJ, Yalcin M, Aljada A, Dyskin E, Davis FB, Lin HY, Davis PJ (2008) Tetraiodothyroacetic acid, a small molecule integrin ligand, blocks angiogenesis induced by vascular endothelial growth factor and basic fibroblast growth factor. Angiogenesis 11:183–190

21. Lin HY, Landersdorfer CB, London D, Meng R, Lim CU, Lin C, Lin S, Tang HY, Brown D, Van Scoy B, Kulawy R, Queimado L, Drusano GL, Louie A, Davis FB, Mousa SA, Davis PJ (2011) Pharmacodynamic modeling of anti-cancer activity of tetraiodothyroacetic acid in a perfused cell culture system. PLoS Comput Biol 7:e1001073
22. Yalcin M, Bharali DJ, Lansing L, Dyskin E, Mousa SS, Hercbergs A, Davis FB, Davis PJ, Mousa SA (2009) Tetraidothyroacetic acid (tetrac) and tetrac nanoparticles inhibit growth of human renal cell carcinoma xenografts. Anticancer Res 29:3825–3831
23. Yalcin M, Dyskin E, Lansing L, Bharali DJ, Mousa SS, Bridoux A, Hercbergs AH, Lin HY, Davis FB, Glinsky GV, Glinskii A, Ma J, Davis PJ, Mousa SA (2010) Tetraiodothyroacetic acid (tetrac) and nanoparticulate tetrac arrest growth of medullary carcinoma of the thyroid. J Clin Endocrinol Metab 95:1972–1980
24. Yalcin M, Bharali DJ, Dyskin E, Dier E, Lansing L, Mousa SS, Davis FB, Davis PJ, Mousa SA (2010) Tetraiodothyroacetic acid and tetraiodothyroacetic acid nanoparticle effectively inhibit the growth of human follicular thyroid cell carcinoma. Thyroid 20:281–286
25. Mousa SA, Yalcin M, Bharali DJ, Meng R, Tang HY, Lin HY, Davis FB, Davis PJ (2012) Tetraiodothyroacetic acid and its nanoformulation inhibit thyroid hormone stimulation of non-small cell lung cancer cells in vitro and its growth in xenografts. Lung Cancer 76:39–45
26. Bharali DJ, Y. M., Davis PJ, Mousa SA (2013) Tetraiodothyroacetic acid (Tetrac) conjugated PLGA nanoparticles: a nanomedicine approach to treat drug-resistant breast cancer. Nanomedicine (Lond.) Feb 28 (Epub ahead of print). doi:10.2217/nnm.12.200
27. Bharali DJ, Klejbor I, Stachowiak EK, Dutta P, Roy I, Kaur N, Bergey EJ, Prasad PN, Stachowiak MK (2005) Organically modified silica nanoparticles: a nonviral vector for in vivo gene delivery and expression in the brain. Proc Natl Acad Sci U S A 102:11539–11544
28. Roy I, Ohulchanskyy TY, Bharali DJ, Pudavar HE, Mistretta RA, Kaur N, Prasad PN (2005) Optical tracking of organically modified silica nanoparticles as DNA carriers: a nonviral, nanomedicine approach for gene delivery. Proc Natl Acad Sci U S A 102:279–284
29. Hood JD, Bednarski M, Frausto R, Guccione S, Reisfeld RA, Xiang R, Cheresh DA (2002) Tumor regression by targeted gene delivery to the neovasculature. Science 296:2404–2407
30. Kim WJ, Yockman JW, Lee M, Jeong JH, Kim YH, Kim SW (2005) Soluble Flt-1 gene delivery using PEI-g-PEG-RGD conjugate for anti-angiogenesis. J Control Release 106:224–234
31. Zhan C, Meng Q, Li Q, Feng L, Zhu J, Lu W (2012) Cyclic RGD-polyethylene glycol-polyethylenimine for intracranial glioblastoma-targeted gene delivery. Chem Asian J 7:91–96
32. Leng A, Yang J, Liu T, Cui J, Li XH, Zhu Y, Xiong T, Chen Y (2011) Nanoparticle-delivered VEGF-silencing cassette and suicide gene expression cassettes inhibit colon carcinoma growth in vitro and in vivo. Tumour Biol 32:1103–1111
33. Lu ZX, Liu LT, Qi XR (2011) Development of small interfering RNA delivery system using PEI-PEG-APRPG polymer for antiangiogenic vascular endothelial growth factor tumor-targeted therapy. Int J Nanomedicine 6:1661–1673
34. Hadj-Slimane R, Lepelletier Y, Lopez N, Garbay C, Raynaud F (2007) Short interfering RNA (siRNA), a novel therapeutic tool acting on angiogenesis. Biochimie 89:1234–1244
35. Li YH, Shi QS, Du J, Jin LF, Du LF, Liu PF, Duan YR (2013) Targeted delivery of biodegradable nanoparticles with ultrasound-targeted microbubble destruction-mediated hVEGF-siRNA transfection in human PC-3 cells in vitro. Int J Mol Med 31:163–171
36. Schiffelers RM, Ansari A, Xu J, Zhou Q, Tang Q, Storm G, Molema G, Lu PY, Scaria PV, Woodle MC (2004) Cancer siRNA therapy by tumor selective delivery with ligand-targeted sterically stabilized nanoparticle. Nucleic Acids Res 32:e149
37. Pille JY, Li H, Blot E, Bertrand JR, Pritchard LL, Opolon P, Maksimenko A, Lu H, Vannier JP, Soria J, Malvy C, Soria C (2006) Intravenous delivery of anti-RhoA small interfering RNA loaded in nanoparticles of chitosan in mice: safety and efficacy in xenografted aggressive breast cancer. Hum Gene Ther 17:1019–1026
38. Liu XQ, Xiong MH, Shu XT, Tang RZ, Wang J (2012) Therapeutic delivery of siRNA silencing HIF-1 alpha with micellar nanoparticles inhibits hypoxic tumor growth. Mol Pharm 9:2863–2874

Chapter 15
Biomarkers of Response and Resistance to Anti-angiogenic Treatment

Dan G. Duda

Abstract Over the last decade, anti-angiogenic therapy for cancer has become increasingly used as a standard therapeutic approach for many cancer types. It has also become a standard of care for certain eye diseases. Yet, despite the use of molecularly targeted drugs with well-defined targets, there are currently no validated biological markers (or biomarkers) for appropriately selecting patients for anti-angiogenic therapy. Nor are there biomarkers identifying escape pathways that should be targeted after tumors develop resistance to a given anti-angiogenic drug. A number of potential systemic, circulating, tissue and imaging biomarkers have emerged from recently completed phase I/II/III studies of anti-angiogenic agents. Some of these are measured at baseline while others are measured during treatment – and all are mechanistically based. Some of these biomarkers may be pharmacodynamic, for example the increase in circulating VEGF and placental growth factor (PlGF). Others have potential for predicting clinical benefit or identifying the escape pathways, for example stromal-derived factor 1 alpha (SDF1α), interleukin 6 (IL-6) or angiopoietin 2 (Ang-2). Biomarkers of anti-angiogenesis may be disease and/or agent specific, and all of them need to be validated prospectively. In this chapter, I discuss the current challenges in establishing biomarkers of anti-angiogenic treatment. I also define the molecular and cellular biomarkers measured in blood circulation and tumor tissues, discuss their advantages and disadvantages, and comment on the future opportunities for validating biomarkers of anti-angiogenic therapy.

Conflict of Interests Disclosure: D.G.D. does not have any direct financial relation with the commercial identities mentioned in this manuscript.

D.G. Duda, DMD, PhD (✉)
Steele Laboratory for Tumor Biology, Department of Radiation Oncology, Massachusetts General Hospital, Harvard Medical School, Boston, MA, USA
e-mail: duda@steele.mgh.harvard.edu

S.A. Mousa and P.J. Davis (eds.), *Angiogenesis Modulations in Health and Disease:* 181
Practical Applications of Pro- and Anti-angiogenesis Targets,
DOI 10.1007/978-94-007-6467-5_15, © Springer Science+Business Media Dordrecht 2013

Targeting New Blood Vessel Formation in Malignant Solid Tumors: Progress and Challenges

Tumors acquire blood vessels by co-option of neighboring vessels, from sprouting or intussusceptive microvascular growth, and by vasculogenesis from endothelial precursor cells [1]. In most solid tumors the newly formed vessels are plagued by structural and functional abnormalities owing to the sustained and excessive exposure to angiogenic factors produced by the growing tumor [2]. However abnormal, these new vessels allow tumor expansion at early stages of carcinogenesis and progression from *in situ* lesions to locally invasive, and eventually to metastatic tumors. Over the past four decades, a large body of preclinical evidence confirmed experimentally the hypothesis that tumor progression can be arrested by anti-angiogenesis [3, 4]. In the clinical setting, the United States Food and Drug Administration has approved over the last decade seven other anti-angiogenic agents for cancer treatment and three anti-angiogenic agents for wet age-related macula degeneration therapy (Table 15.1). A large number of other anti-angiogenic agents are in late phases of clinical development (randomized phase III clinical trials).

All the approved anti-angiogenic drugs target VEGF signaling. Some are blocking the ligand, VEGF; e.g., bevacizumab, aflibercept (Zaltrap®/Eylea®, Sanofi-Aventis, Paris, France and Regeneron Pharmaceuticals, Tarrytown, NY, USA), ranibizumab

Table 15.1 Anti-angiogenic drugs approved by the United States Food and Drug Administration (2004–2013)

Anti-VEGF drug	Approved indication
Bevacizumab	Metastatic colorectal cancer (with chemotherapy)
	Metastatic non-squamous non-small cell lung cancer (with chemotherapy)
	Metastatic breast cancer (with chemotherapy)
	Recurrent glioblastoma (monotherapy)
	Metastatic renal cell carcinoma (with IFNα)
Sunitinib	Metastatic renal cell carcinoma (monotherapy)
	Gastrointestinal stromal tumors (monotherapy)
	Pancreatic neuroendocrine tumors (monotherapy)
Sorafenib	Metastatic renal cell carcinoma
	Unresectable hepatocellular carcinoma
	Advanced medullary thyroid cancer
Pazopanib	Metastatic renal cell carcinoma
	Advanced soft tissue sarcoma
Vandetanib	Advanced medullary thyroid cancer
Axitinib	Advanced renal cell carcinoma
Regoranfenib	Metastatic colorectal cancer
Aflibercept	Metastatic colorectal cancer (with chemotherapy)
	Wet age-related macula degeneration
Pegaptanib	*Wet age-related macula degeneration*
Ranibizumab	*Wet age-related macula degeneration*

Reproduced with permission from ref. [5]

(Lucentis®, Genentech, South San Francisco, CA, USA), and pegaptanib (Macugen®, OSI Pharmaceuticals, Long Island, NY, USA). Others are inhibiting the activity of the VEGF tyrosine kinase receptors (VEGFR-1, VEGFR-2), e.g., sorafenib (Nexavar®, Bayer Healthcare Pharmaceuticals, Leverkusen, Germany and Onyx Pharmaceuticals, South San Francisco, CA, USA), sunitinib (Sutent®) and axitinib (Inlyta®, Pfizer Inc., New York, NY, USA), pazopanib (Votrient®, GlaxoSmithKline, Brentford, Middlesex, UK), and vandetanib (Zactima®, AstraZeneca Pharmaceuticals, Alderley Park, Cheshire, UK). Anti-VEGF therapy has become a standard of care for metastatic colorectal cancer (in first, second, and third line of treatment), advanced non-small cell lung cancer, renal cell carcinoma, hepatocellular carcinomas, glioblastoma, gastrointestinal stromal tumor (GIST), pancreatic neuroendocrine tumor, and medullary thyroid cancer [6–18] (Table 15.1). Given these developments, anti-angiogenic therapy represents one of the most exciting areas in cancer research and clinical oncology [19–28]. These agents have changed the practice of oncology but stimulated important questions: How do these therapies work in patients? Is their mechanism of action in patients the same as originally envisioned for anti-angiogenic agents? Is it the same as demonstrated in animal models? Could the overall survival benefit be increased beyond a few months? Could we successfully use these agents in the adjuvant setting following surgical resection? Why do some patients develop severe toxicities from anti-angiogenic therapy? Why is the benefit from anti-angiogenic therapies seen only in some patients? How do we select these patients or the most appropriate therapy? Why do tumors stop responding to anti-angiogenic therapy? What new pathways should be targeted to optimize the response and prolong the duration of response and survival without increasing toxic effects? How do we tailor these new therapies to individual patients? How do we schedule them with contemporary and future therapeutics? The answers to these fundamental questions are not fully known for the approved anti-angiogenic drugs and will be critical in choosing the appropriate agent(s) and to determine their optimal dose and schedule. Only validation of pharmacodynamic, prognostic, predictive, and surrogate biomarkers (see Box 15.1 for definitions) can help address these questions.

Angiogenic Pathways in Solid Cancers

Over four decades of research on angiogenesis have unraveled many of the underpinnings of tumor angiogenesis in general and the VEGF pathway in particular [1, 3, 4, 33, 34]. VEGF is a key pro-angiogenic molecule in developmental neovascularization as well as in physiological and pathological angiogenesis [34–37]. VEGF exerts its effect by binding two tyrosine kinase (TK) receptors VEGFR-1 (FLT1) and VEGFR-2 (KDR) as well as the non-TK receptors neuropilin 1 (NRP-1) and NRP-2 [1]. Of these, VEGF interaction with VEGFR-2 is thought to convey most of the critical pro-angiogenic signals [23]. However, VEGF interaction with VEGFR-1 and NRP-1 in cancer cells or in non-endothelial stromal cells (e.g., in myeloid cells such as macrophages or Gr-1+ myeloid cells) may be critical for the growth of tumors that depend on this pathway for survival and, through indirect

Box 15. 1 Defining Biomarkers

According to current US Food and Drug Administration draft guidance, a biological marker, or biomarker, is defined as a characteristic that is objectively measured and evaluated as an indicator of normal biologic processes, pathogenic processes, or biological responses to a therapeutic intervention [29]. Such characteristics may include genetic differences, either inherited by the individual in the germline or residing in the tumor, or both; changes in RNA, protein, or metabolite levels as a consequence of the disease or of the therapeutic process; changes in physiologic or systemic parameters, such as blood pressure; or anatomical parameters, such as tumor growth, stasis, or shrinkage [29].

When considering biomarker research, it is important to be aware of the different types of biomarkers that can be identified and the limitations posed by certain types of studies. Biomarkers that can be used before treatment include *prognostic* markers, which predict patient outcome regardless of treatment, and *predictive* markers, which provide information about the effect of a specific therapeutic intervention, usually compared to another. The majority of clinical studies of anti-angiogenic agents to date have identified mainly potential prognostic rather than predictive biomarkers because the studies were either too small to show a statistical difference between treatment arms with respect to the biomarker, or because they were early stage trials that included only one treatment arm [30]. This situation is changing as many larger trials of anti-angiogenic cancer therapies are now incorporating pre-planned biomarker analyses. *Pharmacodynamic* biomarkers are used during treatment to monitor its course, and/or to detect resistance or drug toxicity. Ideally, both predictive and pharmacodynamic biomarkers should reflect modulation of an identified biological target of the therapy in question. While this requirement is more straightforward for agents that target oncogenic pathways in cancer cells, it may be difficult to attain for biomarkers of anti-angiogenic agents, since their exact mechanisms of action are not yet well defined [31].

In addition to their functional characteristics, biomarkers should be robust, reliable, reproducible, feasible for use in a clinical setting, and carefully validated as to specificity and sensitivity. The US Food and Drug Administration is currently taking an active role in setting standards for biomarker development, and pharmacogenomic biomarkers have been incorporated into drug labels for multiple oncologic therapies, both targeted and untargeted. Such guidance will prove increasingly important as cancer treatment becomes more personalized and as new therapies are developed that are designed to hit ever-more specific targets in tumorigenic, angiogenic, and genetic pathways.

Reproduced with permission from Duda, Angiogenesis Foundation e-Publication 2011 [32].

mechanisms, to angiogenesis in tumors [1]. During development and in physiological conditions, the effects of VEGF are finely tuned and counterbalanced by anti-angiogenic molecules such as the soluble form of VEGFR-1 (sVEGFR-1/sFLT1) or thrombospondins (TSP-1 and -2), which ensures stabilization and maturation of the vasculature [38]. In tumors, oncogene or hypoxia-driven VEGF overexpression leads to dysregulated angiogenesis and an abnormal vasculature [1]. Here, the balance is tipped toward pro-angiogenesis. In genetic models in mice, overexpression of VEGFR-1 and sVEGFR-1 led to a "normalization" of the tumor vasculature [39]. Conversely, overexpression of sVEGFR-1 may lead to hypertension (e.g., preeclampsia) or defects in developmental angiogenesis [40–42]. Beyond VEGF, other VEGF family members can bind the VEGFRs and participate in angiogenesis: PlGF, VEGF-B, VEGF-C and VEGF-D. The role of VEGFR-1 and its more selective ligand placental growth factor (PlGF) is currently unclear, but may be particularly important in certain malignancies [1]. Similarly, VEGF-C and VEGF-D might play a role during new blood vessel formation [43]. First, they can bind to VEGFR-2 and second, their cognate receptor VEGFR-3 is expressed on "tip" cells (specialized endothelial cells responsible for vessel sprouting) [1, 43]. In addition, other angiogenesis modulators (positive or negative) may affect the angiogenic balance, e.g., the pro-angiogenic molecules basic fibroblast growth factor (bFGF or FGF-2), angiopoietin 1 (Ang-1), Ang-2 and endoglin or the endogenous angiogenesis inhibitors TSP-1 and TSP-2 [1]. These angiogenic molecules are produced by the cancer cells and by the stromal cells alike. The latter include activated tumor-activated fibroblasts and bone marrow-derived cells recruited by the tumor – most notably tumor infiltrating macrophages, neutrophils, and myeloid-derived suppressor cells [44–46].

Could we exploit all this knowledge for biomarker discovery? The answer is likely yes, provided that biomarker studies will be biology-driven and prospectively validated [29–32, 47]. The systemic and imaging biomarkers may also play a crucial role in discovery of biomarkers for anti-angiogenic therapy, and are discussed in detail elsewhere [31, 48, 49]. Here, I will discuss in the following sections the candidate biomarkers belonging to the VEGF family and to other angiogenic pathways, and the cellular biomarkers.

Molecular and Cellular Biomarker Candidates for Anti-angiogenic Therapy

Tissue-based biomarkers are ideal because they reflect the changes occurring in a tumor during treatment, but obtaining biopsies is difficult owing to the invasive nature of the procedure. Circulating molecular and cellular biomarkers found in blood are a minimally invasive alternative that can be used repeatedly over the course of treatment with an anti-angiogenic agent. Whereas changes in blood circulation may reflect the systemic effects of anti-VEGF therapy, the impact of these changes on tumor response or escape remains unclear and will need to be established in mechanistic studies in preclinical models [32].

VEGF Family Members as Circulating Biomarkers

VEGF expression is usually elevated both in the tumors as well as in the cancer patients' circulation and is often an indicator of poor prognosis. All the anti-angiogenic drugs that have received or are pending approval from the US Food and Drug Administration target VEGF signaling – either by blocking the ligand (bevacizumab, aflibercept) or by inhibiting the tyrosine kinase receptors (sorafenib, sunitinib, vandetanib, pazopanib, axitinib and regorafenib). Thus, the natural choice for a biomarker has been VEGF itself. However, to date the results have been highly inconsistent [31]. High VEGF levels are almost invariably associated with poor outcomes in correlative studies [50], which is indicative of its prognostic biomarker value. In some cancers (e.g., breast cancers or HCC) the levels of circulating VEGF in plasma are correlated with outcome of anti-VEGF therapy [31, 50, 51]. However, in other cancers neither the intra-tumoral nor the circulating VEGF is associated with outcome of bevacizumab treatment [52, 53]. For example, a recent meta analysis across four randomized phase III trials of bevacizumab with chemotherapy or immunotherapy in metastatic colorectal cancer, advanced non-small cell lung cancer, and advanced renal cell carcinoma showed that higher baseline levels of circulating VEGF were associated with shortened progression-free survival and overall survival regardless of bevacizumab treatment [54]. This indicates that circulating VEGF levels may be prognostic but not predictive biomarkers for bevacizumab-containing regimens. Moreover, the authors did not find a good correlation between blood circulating VEGF concentration and intra-tumor expression of VEGF [54]. On the other hand, more recent studies have measured shorter isoforms of VEGF (e.g., $VEGF_{121}$), which do not bind to the extracellular matrix components (i.e., heparin), and have found intriguing correlations with outcome [55]. However, other studies failed to detect a significant correlation for short isoforms of VEGF [56]. Thus, the clinical significance of circulating or tissue VEGF levels remains to be clarified, as most of the efforts to use VEGF itself as a predictive biomarker have thus far been disappointing. Current ongoing efforts to measure distinct VEGF isoforms or VEGF fragments may yield additional insight and resurrect interest in research on VEGF as predictive biomarker.

Other VEGF Family Members

In addition to VEGF-A (or VEGF), the VEGF family includes VEGF-B, VEGF-C, VEGF-D, and PlGF. These VEGF family members may play a role in tumor angiogenesis [1]. Currently available anti-angiogenic drugs affect these factors in a differential manner (i.e., they are not affected by bevacizumab but are blocked by aflibercept or tyrosine kinase inhibitors) [1]. Of interest, some of these factors have been shown to be up-regulated in response to anti-VEGF therapy both in patients and in preclinical models [31]. The most consistent change has been the increase in circulating levels of plasma PlGF, which has been reported essentially for all anti-VEGF drugs and experimental agents, irrespective of their mechanism of VEGF inhibition [31, 57]. This has led to the hypotheses that (1) PlGF change may have pharmacodynamic biomarker value and (2) that PlGF increase may mediate resistance to anti-VEGF agents that do not block this molecule (e.g., bevacizumab). Both of these hypotheses need to be further validated prospectively. Of interest, the increase in PlGF may be due to systemic effects, as tumor-derived PlGF may actually be decreased after bevacizumab treatment [58]. Similarly, VEGF-C and VEGF-D have been proposed as escape biomarkers for bevacizumab in metastatic colorectal cancer patients in other exploratory studies [59].

Soluble VEGF Receptors

As discussed above, there are three VEGF tyrosine kinase receptors in the plasma membrane, known as VEGFR-1 (FLT1), VEGFR-2 (KDR), and VEGFR-3 (FLT4). In addition to the plasma membrane receptors, soluble receptors are present in blood circulation – as a result of alternative splicing or possibly due to plasma membrane receptor shedding [31]. Of these soluble receptors, sVEGFR-1 has clear biological activity. This has led our group to conduct extensive studies of circulating sVEGFR-1, an endogenous blocker of VEGF and PlGF and a factor linked with "vascular normalization", as biomarker or response to anti-VEGF agents [60]. Our hypothesis has been that circulating plasma sVEGFR-1 is a "negative" biomarker that could be used to predict response to anti-VEGF therapies in cancer. Specifically, we proposed that cancer patients with pre-existing high levels of circulating sVEGFR-1 (i.e., in whom VEGF pathway is endogenously suppressed) are resistant to bevacizumab and other anti-VEGF treatments. Indeed, we have shown in exploratory studies that patients with higher plasma levels of sVEGFR-1 have a poor outcome after treatment with bevacizumab, sunitinib, vandetanib, and cediranib [60–66]. Collectively, these results suggest that anti-VEGF therapy may not have a beneficial effect in patients with high sVEGFR-1 levels. In further support of this, we also found that patients with higher sVEGFR-1 levels in circulation experienced fewer side effects from anti-VEGF treatments [60, 65, 66]. Finally, polymorphisms in the *FLT1* gene that are associated with higher VEGFR-1 expression have also been associated with poor outcome of bevacizumab-containing regimens in phase III studies (see below) [67]. If confirmed in larger studies, plasma sVEGFR-1 may potentially allow stratification of cancer patients to regimens that include anti-VEGF therapy.

Soluble VEGFR-2, which is an abundant protein in human plasma, has also been extensively studied. Multiple studies have shown that anti-VEGFR tyrosine kinase inhibitors but not bevacizumab induce a significant decrease in plasma sVEGFR-2 levels [summarized in Ref. [31]]. The same result has been reported for circulating sVEGFR-3 (i.e., a decrease in plasma sVEGFR-3 after treatment with tyrosine kinase inhibitors that block VEGFR-3). The presence of this signature has been associated with improved outcomes in some studies, but its value as a predictive or pharmacodynamic biomarker is currently unknown [30–32].

Other Soluble Plasma Biomarker Candidates

Solule Basement Membrane Components

Collagen IV is one of the main constituents of vascular basement membranes. In glioblastomas, there is an excessive deposition of basement membranes, which more than doubles the thickness of tumor blood vessels compared to normal brain blood vessels [68, 69]. Vascular normalization after anti-VEGF therapy results in normalization of the vascular basement membrane – i.e., a reduction in thickness – as seen in mice and in patients [68–70]. Thus, we tested the hypothesis that proteolytic degradation of these membranes could release soluble collagen IV in blood circulation, and that this biomarker could be used as a measure of therapeutic efficacy. Indeed, we found that recurrent glioblastoma patients who had an increase in plasma collagen IV levels after anti-VEGF therapy had an increase in progression-free survival [71]. If validated, either alone or in combination with imaging biomarkers of vascular normalization, the change in soluble collagen IV may potentially allow an early assessment of drug activity and stratification of glioblastoma patients to anti-VEGF therapies [71].

Inflammatory Factors

In addition to VEGF family members, many biomarker studies have focused on inflammatory cytokines and chemokines because they may exert pro-angiogenic effects either directly or indirectly (via modulation of bone marrow-derived cell recruitment in circulation and infiltration in tumors) (Box 15.2).

A comprehensive study was conducted in patients with advanced non-small cell lung cancer who were treated with vandetanib plus chemotherapy, vandetanib alone, or chemotherapy alone. Interestingly, the patterns of changes in soluble biomarkers in each of the three study arms were distinct [72]. Specifically, an increased risk of disease progression was associated with increases in a different marker in each arm: increased plasma VEGF levels for vandetanib monotherapy versus increases in plasma Interleukin (IL)-8 concentration for combination therapy. IL-8 may act as a VEGF-independent pro-angiogenic pathway [73] and has been associated with

Box 15.2 Inflammatory Molecules and Their Potential Role in Liver Cancer Angiogenesis

Chronic inflammation is a potential precursor and promoter of carcinogenesis in many cancers [45, 78–80]. In many cancers, nuclear factor kappa B (NF-kB) is involved in tumor initiation and progression mediated via STAT3 activation [81–83]. Inflammatory cytokines induced by NF-kB pathway activation might affect angiogenesis directly via endothelial cells, or indirectly by cancer cells or recruitment and/or activation of inflammatory cells [84–91]. Interleukin (IL)-1α has a critical role by recruitment of inflammatory cells [92, 93]. Tumor necrosis factor (TNF)-α can also promote tumor progression by different pathways: direct effect on tumor cells, induction of CXCR4 and stimulation of epithelial – mesenchymal transition [94]. TNF-α promotes cell survival and angiogenesis or induces endothelial cell apoptosis, and vascular disruption and increased permeability. IL-6 is also induced by activation of NF-kB and other transcription factors (C/EPBb and AP-1), and modulates inflammation via IL-6R and gp130. Vascular smooth muscle cells, T lymphocytes and macrophages secrete IL-6 to stimulate immune responses and promote inflammation. IL-6 may also have anti-inflammatory effects by inhibition of TNF-α and IL-1, and activation of IL-1Ra and IL-10. The proliferative and survival effects of IL-6 are mediated by STAT3 [79]. Moreover, IL-8 may have a role in cancer cell invasion [73, 95]. IL-8 can promote tumorigenesis and angiogenesis through CXCR1 and CXCR2, and the Duffy antigen receptor for cytokines, which has no defined intracellular signaling capabilities [96]. Overexpression of VEGF induces the expression of the CXCR4 ligand – stromal cell derived factor 1 alpha (SDF1α) or CXCL12, and SDF1α and CXCR4 may drive cell migration and angiogenesis by VEGF-independent mechanisms [97, 98]. Stem Cell Factor (also known as SCF or Kit-ligand) is a cytokine that binds to the c-Kit receptor (CD117), primarily expressed by early hematopoietic precursors. While c-Kit expression is rarely detectable in the cancer cells, both SCF and c-Kit could be expressed during carcinogenesis, for example in cholangiocarcinomas [99].

Abbreviations: AP-1, activator protein 1; C/EPB, CAAT/enhancer binding-protein; CXCR, C-X-C-chemokine receptor; STAT, signal transducers and activators of transcription.

Adapted with permission from Zhu AX, Duda DG, Sahani DV, Jain RK. HCC and angiogenesis: possible targets and future directions. Nature Reviews Clinical Oncology 2011 [50].

poor prognosis in hepatocellular carcinoma patients treated with sunitinib [61]. Other notable candidates for biomarkers of tumor evasion from anti-VEGF therapy are the stromal-cell-derived factor 1 alpha (SDF1α, also referred to as CXCL12) and IL-6. We have found associations between increased plasma SDF1α after treatment and poor outcome in studies of anti-VEGF agents in recurrent glioblastoma (cediranib), sarcoma (sorafenib), and breast cancer (bevacizumab) patients [62, 65, 74–76]. Moreover, increased plasma SDF1α and plasma IL-6 have been associated with poor outcomes in locally advanced rectal cancer after treatment with bevacizumab and chemoradiation and in advanced hepatocellular carcinoma patients after treatment with sunitinib [61, 77]. These potential resistance biomarkers may drive the design of trials of anti-VEGF agents.

Other Circulating Factors or Soluble Receptors

Finally, recent studies have reported significant changes or associations with outcome for other circulating factors and/or their soluble receptors. Some of the findings have been more consistent, for example the transient decrease in plasma Ang-2 after anti-VEGF therapy [62, 63, 65]. Others appeared to be more agent/disease specific, for example changes and correlations between circulating bFGF, platelet derived growth factor (PDGF)-BB, soluble (s)Tie2, soluble inter-cellular adhesion molecule 1 (sICAM-1), and matrix metalloproteinase (MMP)-2, MMP-9, and MMP-10 [59, 61–63, 65, 72, 100]. All of these biomarkers will require additional study and prospective validation.

Tissue-based Biomarkers

Whenever available – for example when serial biopsies can be performed or when tissues are obtained at surgery or autopsy – tumor specimens have been invaluable for conducting correlative studies and gaining mechanistic insights into the effects of anti-VEGF therapies. These studies have been quite limited because of the invasive and costly nature of these procedures and the difficulty in standardizing immunohistochemical procedures.

As mentioned previously, intra-tumoral levels of VEGF have not been so far shown to predict survival outcome of anti-VEGF therapy [53, 54], although correlations with response rates have been reported [101, 102]. Given the disappointing data reported so far, and considering the limitations of tissue VEGF evaluation, this biomarker does not appear promising.

However, these intriguing results raised critical questions. If neither circulating nor tissue VEGF correlate with outcome of anti-VEGF agents, then what is the mechanism of action that leads to a benefit after treatment with these drugs? While multiple groups are actively exploring various mechanisms involving the vasculature, stroma, immune system, or cancer cells themselves, several emerging data are standing out. Tumor microvascular density has been often evaluated both as a predictive biomarker and as a pharmacodynamic marker of anti-angiogenic therapy with anti-VEGF agents. Indeed, two studies found a decrease in vascular density after bevacizumab treatment in rectal and breast cancer [65, 77, 103]. But other studies did not find a significant change [104]. This effect was associated with increased apoptotic rate in cancer cells, but interestingly, did not change the proliferation rate of cancer cells [103, 104]. One explanation for this paradoxical finding is that the remaining vasculature after anti-VEGF therapy is more "normal" structurally and functionally [2, 105–107]. The association between microvascular density and survival remains unclear, with most studies reporting a lack of correlation [53]

In a study of serial biopsies from rectal cancers, our group has reported that while bevacizumab did not change VEGF or VEGFR expression in the cancer cells, this anti-VEGF treatment decreased PlGF and increased SDF1α and its receptor (CXCR4) expression in the rectal cancer cells [58]. Of interest, increased plasma SDF1α levels during treatment in these patients correlated with distant disease progression pointing toward SDF1α/CXCR4 axis as a potential escape mechanism from anti-VEGF therapy [58, 74].

While enticing, these hypotheses on the mechanism of action of anti-VEGF agents remain to be further confirmed in patients, as our understanding of the dynamics of VEGFR regulation and the interactions between receptor subtypes in tumor tissue is not well enough advanced to allow the use of these levels as biomarkers of therapeutic efficacy.

Finally, genetic studies of tumor samples have also generated mixed results. While establishing the mutational status in various cancers has made a crucial impact on the development and use of anti-cancer agents, e.g., *KRAS* mutation for cetuximab treatment in metastatic colorectal cancer and *BRAF* mutation for vemu-rafenib treatment in melanoma, it has failed so far to impact the development or the use of anti-VEGF drugs. For example, *P53, KRAS,* or *BRAF* mutations in metastatic colorectal cancer did not associate with bevacizumab-chemotherapy treatment outcome in metastatic colo rectal cancer [108]. Many studies have focused on single nucleotide polymorphisms (SNPs) in VEGF family genes as well as other genes [109–112]. Some reports found significant correlations between certain VEGF and VEGFR-2 genes with survival or risk of developing hypertension after bevacizumab treatment in metastatic breast and colorectal cancer [111, 113]. However, these findings have not been yet reproduced by other studies. More recently, SNPs in *FLT1* were shown to associate survival after treatment with bevacizumab-based regimens in two phase III studies in advanced pancreatic adenocarcinoma and metastatic renal cell carcinoma [67]. These SNPs were associated with higher VEGFR-1 expression [67]. These *FLT1* SNPs correlated with a poor outcome, which is in line with the finding that high circulating sVEGFR-1 is associated with poor outcome after anti-VEGF therapy (see above) [60–66]. Also, a consistent finding appears to be the association between SNPs in *CXCR2* and *IL8* genes and outcome after anti-VEGF therapies [109, 110, 112, 114]. Once again, this suggests an important role that inflammatory cytokines and their receptors may play in the outcome of anti-VEGF therapy. These data strongly suggest that SNP evaluation could be used in the future to predict outcome of anti-VEGF therapy. Moreover, the evaluations of gene polymorphisms have the great advantage of being more feasible as they are minimally invasive, less expensive, and do not necessarily require tumor tissue. However, only more extensive investigation and validation of the current lead candidates could potentially provide a biomarker for anti-VEGF therapy.

Challenges, Conclusions and Future Perspective

One major challenge for the interpretation of molecular biomarker studies in general is that a vast amount of data was generated in single arm studies, i.e., in which all patients received the same therapy. This makes the distinction between prognostic and predictive biomarkers impossible. Another challenge is that while bevacizumab and aflibercept are specific inhibitors of VEGF pathways, all the anti-angiogenic tyrosine kinase inhibitors are promiscuous, inhibiting multiple, non-angiogenic tyrosine kinases as well as angiogenic ones [115, 116]. Therefore, it can be difficult to know whether a given biochemical or physiological effect is the result of anti-angiogenic activity or due to effects on other oncogenic targets (e.g., c-Kit inhibition by sunitinib in gastrointestinal stromal tumors or EGFR and RET inhibition by vandetanib in advanced medullary thyroid cancer). Even for bevacizumab/aflibercept studies, the interpretation is confounded by the fact that most studies included concurrent chemotherapeutic drugs, making it difficult to tease out the effects of each type of therapy.

In summary, identifying and validating predictive biomarkers of response and gaining the ability to stratify cancer patients to currently approved anti-angiogenic drugs remains a major priority in oncology. A number of potential biomarkers have emerged from correlative clinical studies and warrant further study in large, randomized trials. Some such trials are now underway and their results will be critical for advancement of this field, not only for biomarker discovery but also for further elucidation of the specific mechanisms of action of these important new therapies.

Acknowledgements The author is grateful to Drs. Rakesh K. Jain, Christopher G. Willett, Tracy T. Batchelor, Andrew X. Zhu and their teams for the contributions to the studies summarized in this chapter. Dr. Duda's research is supported by US National Institutes of Health grants P01-CA080124, R01-CA159258, and Federal Share National Cancer Institute Proton Beam Program Income and by the American Cancer Society grant 120733-RSG-11-073-01-TBG. This chapter uses illustrations and content with permission from several recent reviews, referenced herein [5, 31, 32, 50].

References

1. Carmeliet P, Jain RK (2011) Molecular mechanisms and clinical applications of angiogenesis. Nature 473:298–307
2. Goel S, Duda DG, Xu L et al (2011) Normalization of the vasculature for treatment of cancer and other diseases. Physiol Rev 91:1071–1121
3. Folkman J (1971) Tumor angiogenesis: therapeutic implications. N Engl J Med 285:1182–1186
4. Folkman J (2007) Angiogenesis: an organizing principle for drug discovery? Nat Rev Drug Discov 6:273–286
5. Duda DG (2012) Molecular biomarkers of response to antiangiogenic therapy for cancer. ISRN Cell Biology. doi:10.5402/2012/587259

6. Friedman HS, Prados MD, Wen PY et al (2009) Bevacizumab alone and in combination with irinotecan in recurrent glioblastoma. J Clin Oncol 27:4733–4740

7. Demetri GD, van Oosterom AT, Garrett CR et al (2006) Efficacy and safety of sunitinib in patients with advanced gastrointestinal stromal tumour after failure of imatinib: a randomised controlled trial. Lancet 368:1329–1338

8. Escudier B, Eisen T, Stadler WM et al (2007) Sorafenib in advanced clear-cell renal-cell carcinoma. N Engl J Med 356:125–134

9. Giantonio BJ, Catalano PJ, Meropol NJ et al (2007) Bevacizumab in combination with oxaliplatin, fluorouracil, and leucovorin (FOLFOX4) for previously treated metastatic colorectal cancer: results from the Eastern Cooperative Oncology Group Study E3200. J Clin Oncol 25:1539–1544

10. Hurwitz H, Fehrenbacher L, Novotny W et al (2004) Bevacizumab plus irinotecan, fluorouracil, and leucovorin for metastatic colorectal cancer. N Engl J Med 350:2335–2342

11. Llovet JM, Ricci S, Mazzaferro V et al (2008) Sorafenib in advanced hepatocellular carcinoma. N Engl J Med 359:378–390

12. Motzer RJ, Hutson TE, Tomczak P et al (2007) Sunitinib versus interferon alfa in metastatic renal-cell carcinoma. N Engl J Med 356:115–124

13. Cheng A, Kang Y, Lin D et al (2011) Phase III trial of sunitinib versus sorafenib in advanced hepatocellular carcinoma. J Clin Oncol 29S:(abstr 4000)

14. Rini BI, Halabi S, Rosenberg JE et al (2010) Phase III trial of bevacizumab plus interferon alfa versus interferon alfa monotherapy in patients with metastatic renal cell carcinoma: final results of CALGB 90206. J Clin Oncol 28:2137–2143

15. Sandler A, Gray R, Perry MC et al (2006) Paclitaxel-carboplatin alone or with bevacizumab for non-small-cell lung cancer. N Engl J Med 355:2542–2550

16. Sternberg CN, Davis ID, Mardiak J et al (2010) Pazopanib in locally advanced or metastatic renal cell carcinoma: results of a randomized phase III trial. J Clin Oncol 28:1061–1068

17. Van Cutsem E, Tabernero J, Lakomy R et al (2012) Addition of aflibercept to fluorouracil, leucovorin, and irinotecan improves survival in a phase III randomized trial in patients with metastatic colorectal cancer previously treated with an oxaliplatin-based regimen. J Clin Oncol 30(28):3499–3506

18. Wells SA Jr, Robinson BG, Gagel RF et al (2012) Vandetanib in patients with locally advanced or metastatic medullary thyroid cancer: a randomized, double-blind phase III trial. J Clin Oncol 30:134–141

19. Duda DG, Batchelor TT, Willett CG et al (2007) VEGF-targeted cancer therapy strategies: current progress, hurdles and future prospects. Trends Mol Med 13:223–230

20. Duda DG, Jain RK, Willett CG (2007) Antiangiogenics: the potential role of integrating this novel treatment modality with chemoradiation for solid cancers. J Clin Oncol 25:4033–4042

21. Ellis LM (2003) Antiangiogenic therapy at a crossroads: clinical trial results and future directions. J Clin Oncol 21:281s–283s

22. Eskens FA (2004) Angiogenesis inhibitors in clinical development; where are we now and where are we going? Br J Cancer 90:1–7

23. Ferrara N, Hillan KJ, Gerber HP et al (2004) Discovery and development of bevacizumab, an anti-VEGF antibody for treating cancer. Nat Rev Drug Discov 3:391–400

24. Grothey A, Galanis E (2009) Targeting angiogenesis: progress with anti-VEGF treatment with large molecules. Nat Rev Clin Oncol 6:507–518

25. Heath VL, Bicknell R (2009) Anticancer strategies involving the vasculature. Nat Rev Clin Oncol 6:395–404

26. Jain RK (2005) Antiangiogenic therapy for cancer: current and emerging concepts. Oncology (Williston Park) 19:7–16

27. Jain RK, Duda DG, Clark JW et al (2006) Lessons from phase III clinical trials on anti-VEGF therapy for cancer. Nat Clin Pract Oncol 3:24–40

28. Rosen LS (2005) VEGF-targeted therapy: therapeutic potential and recent advances. Oncologist 10:382–391

29. Biomarkers Definitions Working Group (2001) Biomarkers and surrogate endpoints: preferred definitions and conceptual framework. Clin Pharmacol Ther 69:89–95
30. Murukesh N, Dive C, Jayson GC (2010) Biomarkers of angiogenesis and their role in the development of VEGF inhibitors. Br J Cancer 102:8–18
31. Jain RK, Duda DG, Willett CG et al (2009) Biomarkers of response and resistance to antiangiogenic therapy. Nat Rev Clin Oncol 6:327–338
32. Duda DG (2011) Targeting tumor angiogenesis: biomarkers of angiogenesis and antiangiogenic therapy in cancer. Angiogenesis Foundation:e-publication
33. Dvorak HF (2002) Vascular permeability factor/vascular endothelial growth factor: a critical cytokine in tumor angiogenesis and a potential target for diagnosis and therapy. J Clin Oncol 20:4368–4380
34. Ferrara N, Gerber HP, LeCouter J (2003) The biology of VEGF and its receptors. Nat Med 9:669–676
35. Carmeliet P (2005) Angiogenesis in life, disease and medicine. Nature 438:932–936
36. Carmeliet P, Ferreira V, Breier G et al (1996) Abnormal blood vessel development and lethality in embryos lacking a single VEGF allele. Nature 380:435–439
37. Ferrara N, Carver-Moore K, Chen H et al (1996) Heterozygous embryonic lethality induced by targeted inactivation of the VEGF gene. Nature 380:439–442
38. Jain RK (2003) Molecular regulation of vessel maturation. Nat Med 9:685–693
39. Mazzone M, Dettori D, Oliveira R Leite de et al (2009) Heterozygous deficiency of PHD2 restores tumor oxygenation and inhibits metastasis via endothelial normalization. Cell 136:839–51
40. Chappell JC, Taylor SM, Ferrara N et al (2009) Local guidance of emerging vessel sprouts requires soluble Flt-1. Dev Cell 17:377–386
41. Levine RJ, Maynard SE, Qian C et al (2004) Circulating angiogenic factors and the risk of preeclampsia. N Engl J Med 350:672–683
42. Maynard SE, Min JY, Merchan J et al (2003) Excess placental soluble fms-like tyrosine kinase 1 (sFlt1) may contribute to endothelial dysfunction, hypertension, and proteinuria in preeclampsia. J Clin Invest 111:649–658
43. Tammela T, Zarkada G, Wallgard E et al (2008) Blocking VEGFR-3 suppresses angiogenic sprouting and vascular network formation. Nature 454:656–660
44. Hanahan D, Coussens LM (2012) Accessories to the crime: functions of cells recruited to the tumor microenvironment. Cancer Cell 21:309–322
45. Pollard JW (2004) Tumour-educated macrophages promote tumour progression and metastasis. Nat Rev Cancer 4:71–78
46. Murdoch C, Muthana M, Coffelt SB et al (2008) The role of myeloid cells in the promotion of tumour angiogenesis. Nat Rev Cancer 8:618–631
47. Park JW, Kerbel RS, Kelloff GJ et al (2004) Rationale for biomarkers and surrogate end points in mechanism-driven oncology drug development. Clin Cancer Res 10:3885–3896
48. Collins JM (2005) Imaging and other biomarkers in early clinical studies: one step at a time or re-engineering drug development? J Clin Oncol 23:5417–5419
49. Galbraith SM (2003) Antivascular cancer treatments: imaging biomarkers in pharmaceutical drug development. Br J Radiol 76(1):S83–S86
50. Zhu AX, Duda DG, Sahani DV et al (2011) HCC and angiogenesis: possible targets and future directions. Nat Rev Clin Oncol 8:292–301
51. Miles DW, de Haas SL, Dirix LY et al (2013) Biomarker results from the AVADO phase 3 trial of first-line bevacizumab plus docetaxel for HER2-negative metastatic breast cancer. Br J Cancer. ePub on 19 Feb 2013 as doi:10.1038/bjc.2013.69.
52. Dowlati A, Gray R, Johnson DH et al (2006) Prospective correlative assessment of biomarkers in E4599 randomized phase II/III trial of carboplatin and paclitaxel±bevacizumab in advanced non-small cell lung cancer (NSCLC). J Clin Oncol 24S:7027
53. Jubb AM, Hurwitz HI, Bai W et al (2006) Impact of vascular endothelial growth factor-A expression, thrombospondin-2 expression, and microvessel density on the treatment effect of bevacizumab in metastatic colorectal cancer. J Clin Oncol 24:217–227

54. Bernaards C, Hegde P, Chen D et al (2010) Circulating vascular endothelial growth factor (VEGF) as a biomarker for bevacizumab-based therapy in metastatic colorectal, non-small cell lung, and renal cell cancers: analysis of phase III studies. J Clin Oncol 28:15S:(suppl; abstr 10519)
55. Van Cutsem E, de Haas S, Kang YK et al (2012) Bevacizumab in combination with chemotherapy as first-line therapy in advanced gastric cancer: a biomarker evaluation from the AVAGAST randomized phase III trial. J Clin Oncol 30:2119–2127
56. Jayson GC, de Haas S, Delmar P et al (2011) Evaluation of plasma VEGFA as a potential predictive Pan-tumour biomarker for bevacizumab. Eur J Cancer 47:S96
57. Horowitz NS, Penson RT, Duda DG et al (2011) Safety, efficacy and biomarker exploration in a phase II study of bevacizumab, oxaliplatin and gemcitabine in recurrent Müllerian carcinoma. Clin Ovarian Cancer Other Gynecol Malig 4:26–33
58. Xu L, Duda DG, di Tomaso E et al (2009) Direct evidence that bevacizumab, an anti-VEGF antibody, up-regulates SDF1alpha, CXCR4, CXCL6, and neuropilin 1 in tumors from patients with rectal cancer. Cancer Res 69:7905–7910
59. Lieu CH, Tran HT, Jiang Z et al (2011) The association of alternate VEGF ligands with resistance to anti-VEGF therapy in metastatic colorectal cancer. J Clin Oncol 29S:(abstr 3533)
60. Duda DG, Willett CG, Ancukiewicz M et al (2010) Plasma soluble VEGFR-1 is a potential dual biomarker of response and toxicity for bevacizumab with chemoradiation in locally advanced rectal cancer. Oncologist 15:577–583
61. Zhu AX, Sahani DV, Duda DG et al (2009) Efficacy, safety, and potential biomarkers of sunitinib monotherapy in advanced hepatocellular carcinoma: a phase II study. J Clin Oncol 27:3027–3035
62. Batchelor TT, Duda DG, di Tomaso E et al (2010) Phase II study of cediranib, an oral pan-VEGF receptor tyrosine kinase inhibitor, in patients with recurrent glioblastoma. J Clin Oncol 28:2817–2823
63. Gerstner ER, Emblem KE, Chi AS et al (2012) Effects of cediranib, a VEGF signaling inhibitor, in combination with chemoradiation on tumor blood flow and survival in newly diagnosed glioblastoma. J Clin Oncol 30S:(abstr 2009)
64. Meyerhardt JA, Ancukiewicz M, Abrams TA et al (2012) Phase I study of cetuximab, irinotecan, and vandetanib (ZD6474) as therapy for patients with previously treated metastastic colorectal cancer. PLoS One 7:e38231
65. Tolaney SM, Duda DG, Boucher Y et al (2012) A phase II study of preoperative (preop) bevacizumab (bev) followed by dose-dense (dd) doxorubicin (A)/cyclophosphamide (C)/paclitaxel (T) in combination with bev in HER2-negative operable breast cancer (BC). J Clin Oncol 30:(suppl; abstr 1026)
66. Zhu AX, Ancukiewicz M, Supko JG et al (2013) Efficacy, safety, pharmacokinetics and biomarkers of cediranib monotherapy in advanced hepatocellular carcinoma: a phase II study. Clin Cancer Res 19:1157–1166
67. Lambrechts D, Claes B, Delmar P et al (2012) VEGF pathway genetic variants as biomarkers of treatment outcome with bevacizumab: an analysis of data from the AViTA and AVOREN randomised trials. Lancet Oncol 13(7):724–733
68. Kamoun WS, Ley CD, Farrar CT et al (2009) Edema control by cediranib, a vascular endothelial growth factor receptor-targeted kinase inhibitor, prolongs survival despite persistent brain tumor growth in mice. J Clin Oncol 27:2542–2552
69. Winkler F, Kozin SV, Tong R et al (2004) Kinetics of vascular normalization by VEGFR2 blockade governs brain tumor response to radiation: role of oxygenation, angiopoietin-1 and matrix metalloproteinases. Cancer Cell 6:553–563
70. di Tomaso E, Snuderl M, Kamoun W et al (2011) Glioblastoma recurrence after cediranib therapy in patients: lack of "rebound" revascularization as mode of escape. Cancer Res 71:19–28
71. Sorensen AG, Batchelor TT, Zhang WT et al (2009) A "vascular normalization index" as potential mechanistic biomarker to predict survival after a single dose of cediranib in recurrent glioblastoma patients. Cancer Res 69:5296–5300

72. Hanrahan EO, Lin HY, Kim ES et al (2010) Distinct patterns of cytokine and angiogenic factor modulation and markers of benefit for vandetanib and/or chemotherapy in patients with non-small-cell lung cancer. J Clin Oncol 28:193–201

73. Mizukami Y, Jo WS, Duerr EM et al (2005) Induction of interleukin-8 preserves the angiogenic response in HIF-1alpha-deficient colon cancer cells. Nat Med 11:992–997

74. Duda DG, Kozin S, Kirkpatrick ND et al (2011) CXCL12 (SDF1alpha) – CXCR4/CXCR7 pathway inhibition: an emerging sensitizer for anti-cancer therapies? Clin Cancer Res 17:2074–2080

75. Raut CP, Boucher Y, Duda DG et al (2012) Effects of sorafenib on intra-tumoral interstitial fluid pressure and circulating biomarkers in patients with refractory sarcomas (NCI protocol 6948). PLoS One 7:e26331

76. Batchelor TT, Sorensen AG, di Tomaso E et al (2007) AZD2171, a pan-VEGF receptor tyrosine kinase inhibitor, normalizes tumor vasculature and alleviates edema in glioblastoma patients. Cancer Cell 11:83–95

77. Willett CG, Duda DG, di Tomaso E et al (2009) Efficacy, safety, and biomarkers of neoadjuvant bevacizumab, radiation therapy, and fluorouracil in rectal cancer: a multidisciplinary phase II study. J Clin Oncol 27:3020–3026

78. Coussens LM, Werb Z (2002) Inflammation and cancer. Nature 420:860–867

79. Naugler WE, Karin M (2008) The wolf in sheep's clothing: the role of interleukin-6 in immunity, inflammation and cancer. Trends Mol Med 14:109–119

80. Mantovani A, Allavena P, Sica A et al (2008) Cancer-related inflammation. Nature 454:436–444

81. Pikarsky E, Porat RM, Stein I et al (2004) NF-kappaB functions as a tumour promoter in inflammation-associated cancer. Nature 431:461–466

82. He G, Yu GY, Temkin V et al (2010) Hepatocyte IKKbeta/NF-kappaB inhibits tumor promotion and progression by preventing oxidative stress-driven STAT3 activation. Cancer Cell 17:286–297

83. Park EJ, Lee JH, Yu GY et al (2010) Dietary and genetic obesity promote liver inflammation and tumorigenesis by enhancing IL-6 and TNF expression. Cell 140:197–208

84. Zhang W, Zhu XD, Sun HC et al (2010) Depletion of tumor-associated macrophages enhances the effect of sorafenib in metastatic liver cancer models by antimetastatic and antiangiogenic effects. Clin Cancer Res 16:3420–3430

85. Hamsa TP, Kuttan G (2012) Antiangiogenic activity of berberine is mediated through the downregulation of hypoxia-inducible factor-1, VEGF, and proinflammatory mediators. Drug Chem Toxicol 35:57–70

86. Zhao JD, Liu J, Ren ZG et al (2010) Maintenance of Sorafenib following combined therapy of three-dimensional conformal radiation therapy/intensity-modulated radiation therapy and transcatheter arterial chemoembolization in patients with locally advanced hepatocellular carcinoma: a phase I/II study. Radiat Oncol 5:12

87. Rhode J, Fogoros S, Zick S et al (2007) Ginger inhibits cell growth and modulates angiogenic factors in ovarian cancer cells. BMC Complement Altern Med 7:44

88. Wu M, Huang C, Li X et al (2008) LRRC4 inhibits glioblastoma cell proliferation, migration, and angiogenesis by downregulating pleiotropic cytokine expression and responses. J Cell Physiol 214:65–74

89. Veschini L, Belloni D, Foglieni C et al (2007) Hypoxia-inducible transcription factor-1 alpha determines sensitivity of endothelial cells to the proteosome inhibitor bortezomib. Blood 109:2565–2570

90. Belakavadi M, Salimath BP (2005) Mechanism of inhibition of ascites tumor growth in mice by curcumin is mediated by NF-kB and caspase activated DNase. Mol Cell Biochem 273:57–67

91. Shibata A, Nagaya T, Imai T et al (2002) Inhibition of NF-kappaB activity decreases the VEGF mRNA expression in MDA-MB-231 breast cancer cells. Breast Cancer Res Treat 73:237–243

92. Sakurai T, He G, Matsuzawa A et al (2008) Hepatocyte necrosis induced by oxidative stress and IL-1 alpha release mediate carcinogen-induced compensatory proliferation and liver tumorigenesis. Cancer Cell 14:156–165
93. Carmi Y, Voronov E, Dotan S et al (2009) The role of macrophage-derived IL-1 in induction and maintenance of angiogenesis. J Immunol 183:4705–4714
94. Germano G, Allavena P, Mantovani A (2008) Cytokines as a key component of cancer-related inflammation. Cytokine 43:374–379
95. Kubo F, Ueno S, Hiwatashi K et al (2005) Interleukin 8 in human hepatocellular carcinoma correlates with cancer cell invasion of vessels but not with tumor angiogenesis. Ann Surg Oncol 12:800–807
96. Brat DJ, Bellail AC, Van Meir EG (2005) The role of interleukin-8 and its receptors in gliomagenesis and tumoral angiogenesis. Neuro Oncol 7:122–133
97. Grunewald M, Avraham I, Dor Y et al (2006) VEGF-induced adult neovascularization: recruitment, retention, and role of accessory cells. Cell 124:175–189
98. Li W, Gomez E, Zhang Z (2007) Immunohistochemical expression of stromal cell-derived factor-1 (SDF-1) and CXCR4 ligand receptor system in hepatocellular carcinoma. J Exp Clin Cancer Res 26:527–533
99. Mansuroglu T, Ramadori P, Dudas J et al (2009) Expression of stem cell factor and its receptor c-Kit during the development of intrahepatic cholangiocarcinoma. Lab Invest 89:562–574
100. Kopetz S, Hoff PM, Morris JS et al (2010) Phase II trial of infusional fluorouracil, irinotecan, and bevacizumab for metastatic colorectal cancer: efficacy and circulating angiogenic biomarkers associated with therapeutic resistance. J Clin Oncol 28:453–459
101. Foernzler D, Delmar P, Kockx M et al (2010) Tumor tissue based biomarker analysis in NO16966: a randomized phase III study of first-line bevacizumab in combination with oxaliplatin-based chemotherapy in patients with mCR. In: 2010 Gastrointestinal cancers symposium proceedings: Abstract 374
102. Yang SX, Steinberg SM, Nguyen D et al (2008) Gene expression profile and angiogenic marker correlates with response to neoadjuvant bevacizumab followed by bevacizumab plus chemotherapy in breast cancer. Clin Cancer Res 14:5893–5899
103. Willett CG, Boucher Y, Duda DG et al (2005) Surrogate markers for antiangiogenic therapy and dose-limiting toxicities for bevacizumab with radiation and chemotherapy: continued experience of a phase I trial in rectal cancer patients. J Clin Oncol 23:8136–8139
104. Wedam SB, Low JA, Yang SX et al (2006) Antiangiogenic and antitumor effects of bevacizumab in patients with inflammatory and locally advanced breast cancer. J Clin Oncol 24:769–777
105. Jain RK (2001) Normalizing tumor vasculature with anti-angiogenic therapy: a new paradigm for combination therapy. Nat Med 7:987–989
106. Jain RK (2005) Normalization of tumor vasculature: an emerging concept in antiangiogenic therapy. Science 307:58–62
107. Jain RK (2008) Taming vessels to treat cancer. Sci Am 298:56–63
108. Ince WL, Jubb AM, Holden SN et al (2005) Association of k-ras, b-raf, and p53 status with the treatment effect of bevacizumab. J Natl Cancer Inst 97:981–989
109. Gerger A, El-Khoueiry A, Zhang W et al (2011) Pharmacogenetic angiogenesis profiling for first-line bevacizumab plus oxaliplatin-based chemotherapy in patients with metastatic colorectal cancer. Clin Cancer Res 17:5783–5792
110. lo Giudice L, Di Salvatore M, Astone A et al (2010) Polymorphisms in VEGF, eNOS, COX-2, and IL-8 as predictive markers of response to bevacizumab. J Clin Oncol 28S:(suppl; abstr e13502)
111. Loupakis F, Ruzzo A, Salvatore L et al (2011) Retrospective exploratory analysis of VEGF polymorphisms in the prediction of benefit from first-line FOLFIRI plus bevacizumab in metastatic colorectal cancer. BMC Cancer 11:247

112. Schultheis AM, Lurje G, Rhodes KE et al (2008) Polymorphisms and clinical outcome in recurrent ovarian cancer treated with cyclophosphamide and bevacizumab. Clin Cancer Res 14:7554–7563
113. Schneider BP, Wang M, Radovich M et al (2008) Association of vascular endothelial growth factor and vascular endothelial growth factor receptor-2 genetic polymorphisms with outcome in a trial of paclitaxel compared with paclitaxel plus bevacizumab in advanced breast cancer: ECOG 2100. J Clin Oncol 26:4672–4678
114. Zhang WW, Cortes JE, Yao H et al (2009) Predictors of primary imatinib resistance in chronic myelogenous leukemia are distinct from those in secondary imatinib resistance. J Clin Oncol 27:3642–3649
115. Duda DG, Ancukiewicz M, Jain RK (2010) Biomarkers of antiangiogenic therapy: how do we move from candidate biomarkers to valid biomarkers? J Clin Oncol 28:183–185
116. Zhu AX, Duda DG, Sahani DV et al (2009) Development of sunitinib in hepatocellular carcinoma: rationale, early clinical experience and correlative studies. Cancer J 15:263–268

Chapter 16
Speculations on New Directions in Which Angiogenesis May Proceed

Shaker A. Mousa and Paul J. Davis

Abstract Angiogenesis regulation is a function of pro-angiogenesis and anti-angiogenesis factors and processes. These include growth factors, cell surface receptors for these factors on blood vessel cells, modulatory crosstalk between growth factor receptors and other cell surface proteins such as integrin $\alpha v\beta 3$, and transduction of growth factor signals into a modified angiogenesis process. In this chapter we consider areas of blood vessel growth control in which we can expect new information in the near future.

Regulation of angiogenesis is a function of pro-angiogenesis and anti-angiogenesis growth factors, cell surface receptors for these factors on blood vessel cells, modulatory crosstalk between growth factor receptors and other cell surface proteins such as integrin $\alpha v\beta 3$, and transduction of growth factor signals into a modified angiogenesis process. These factors and processes have been emphasized in preceding chapters. Our knowledge of the molecular basis of angiogenesis remains incomplete, and we will briefly consider here areas of control of blood vessel growth in which we can expect new information in the foreseeable future.

S.A. Mousa (✉)
The Pharmaceutical Research Institute at Albany College of Pharmacy
and Health Sciences, Rensselaer, NY, USA
e-mail: shaker.mousa@acphs.edu

P.J. Davis
The Pharmaceutical Research Institute at Albany College of Pharmacy
and Health Sciences, Rensselaer, NY, USA

Department of Medicine, Albany Medical Center, Albany, NY, USA
e-mail: pdavis.ordwayst@gmail.com

S.A. Mousa and P.J. Davis (eds.), *Angiogenesis Modulations in Health and Disease:*
Practical Applications of Pro- and Anti-angiogenesis Targets,
DOI 10.1007/978-94-007-6467-5_16, © Springer Science+Business Media Dordrecht 2013

Novel Vascular Growth Factors, Receptors, and Angiogenesis Signal Transduction

Additional protein growth factors or isoforms of growth factors that modulate angiogenesis are likely to exist. A prototype here is the VEGF family that now includes isoforms VEGF-A through VEGF-E and placental growth factor (PlGF) [1]. Previously unrecognized factors or isoforms may be sought in vascular malformations or organ remodeling after induction of ischemic damage. Further, well-studied naturally occurring products previously unrecognized to have pro-angiogenesis activity may indeed have such activity. Recent examples are melatonin [2] and lipocalin 2 (Lcn2; NGAL) [3] that affect VEGF production and the contribution of testosterone to post-myocardial infarction revascularization in the heart [4]. Erythropoietin is pro-angiogenic [5].

What are the receptor sites for these factors now understood to have pro-angiogenesis actions? In the case of melatonin, are the binding sites involved in angiogenesis the G protein-coupled receptors (MT1, MT2) linked to classical melatonin activities, or are they novel sites such as we describe below on an integrin?

The existence of crosstalk between integrin $\alpha v\beta 3$ and adjacent VEGF and bFGF receptors has been widely appreciated. It has been assumed that such communication at the cell surface or in or immediately beneath the plasma membrane is a function of binding of extracellular matrix proteins by the integrin or communication from within the cell (inside-out messaging). We have emphasized that $\alpha v\beta 3$ has specific, high-affinity receptors for small molecules, for example, for thyroid hormone [6]. The thyroid hormone receptor on the integrin affects the interactions of VEGF and bFGF with their respective cell surface receptors. Depending upon the nature of the thyroid hormone analogue bound to the receptor, the latter may generate pro- or anti-angiogenesis signals. $\alpha v\beta 3$ also bears receptors for testosterone and for resveratrol. Is the testosterone receptor on the integrin the site of initiation of revascularization in the heart described above? Resveratrol is an anti-angiogenesis stilbene, decreasing VEGF production in the setting of human peritoneal mesothelial cell-dependent angiogenesis [7], whereas several other stilbenes are pro-angiogenesis. Is this the initiation site for actions of resveratrol analogues on blood vessel formation? Via a nongenomic mechanism, estrogen has a variety of effects on endothelial cells and angiogenesis [8]. X-ray crystallographic modeling of integrin $\alpha v\beta 3$ suggests the presence of an estrogen-binding domain [9] that would be a candidate initiation site for pro-angiogenesis action of estrogen.

The transduction of signals from traditional pro- and anti-angiogenesis factors expressed at their receptors has been well-studied. Signal transduction has also been defined in the examples of certain small molecules that, via integrin $\alpha v\beta 3$, modulate angiogenesis and has exhibited novel features. The thyroid hormone receptor on $\alpha v\beta 3$ has two binding domains, one of which activates mitogen-activated protein kinase (MAPK; extracellular-regulated kinase 1/2) and the other, phosphatidylinositol 3-kinase (PI3K) (see Chap. 4). The domains distinguish among thyroid hormone analogues to induce either pro- or anti-angiogenesis. Downstream, these pathways have discrete effects on specific gene transcription, for example, on *thrombospondin 1* gene expression [10], introducing an additional and novel mechanism by which thyroid hormone can affect neovascularization. The activation of PI3K by thyroid hormone—3, 5, 3′-triiodo-L-thyronine, T_3, but not L-thyroxine, T_4—at its $\alpha v \beta 3$ receptor leads to transcription of the *hypoxia-inducible factor-1α (HIF-1α)* gene [11] that is relevant to angiogenesis (see below). We would propose that this signal transduction model may be relevant to other small molecule receptor sites on the integrin that relate to angiogenesis. The $\alpha v \beta 3$-MAPK signal transduction pathway is also utilized by non-neuronal nicotinic acetylcholine receptors (AChRs), specifically, $\alpha 7$ AChRs, to stimulate angiogenesis (see Chap. 6). Little is known about possible interactions of thyroid hormone and AChRs, but an insight into the complexity of the regulation of angiogenesis is that nicotine, acting via an AChR, increases the generation of T_3 in brain via deiodination of T_4 (deiodinase 2, D2) [12]. Thus, pro-angiogenesis activity of nicotine may involve thyroid hormone.

Hypoxia is a stimulus to angiogenesis and the mechanism is HIF-1α-dependent. Factors identified in the future to stimulate expression of the *HIF-1α* gene of course may be pro-angiogenesis. Melatonin, lipocalin 2 and testosterone are already known to induce neovascularization, as noted above; these factors are also recognized to cause tissue accumulation of HIF-1α in the settings of malignancy or nonmalignant tissue remodeling. A complex nongenomic intracellular signal transduction pathway links estradiol to induction of *HIF-1α* gene expression, and presumably to angiogenesis, in breast cancer cells [13]

The inflammatory state is associated with new blood vessel formation and is also likely to be a source in the future of newly identified pro-angiogenesis factors. Inflammatory cytokines such as interleukin-8 (IL-8) in the context of cancer [14] and IL-18 in rheumatoid arthritis [15] are pro-angiogenesis. Certain chemokine ligands induce intra-tumoral angiogenesis, e.g., CCL3 and its specific receptor, CCR5 [16].

The foregoing discussion identifies only a few areas of blood vessel biology in which novel pro-angiogenesis factors may be sought. What is clear is that there already is known a plethora of such factors. The potency of each is variable in the settings in which they have been described and apparent importance of new angiogenesis factors may certainly depend upon the experimental or clinical contexts in which they have been described. Of course, net angiogenesis in a given tissue or model setting is the algebraic sum of pro- and anti-angiogenesis substances present. Factors such as interstitial pressure in the intact tumor may modify the expression of the algebraic sum, and thus assay models, e.g., subcutaneous vs. orthotopic xenografts, provide important supplemental information about actions of newly identified individual angiogenesis factors and combinations of factors. Pharmacologic reduction of the permeability of vessels serving tumors reduces interstitial pressure and has been shown to improve tumor delivery of low molecular weight anti-cancer agents and of smaller nanoparticles (12 nm) [17].

Micro RNA and Angiogenesis Modulation

Recent reports suggested that micro RNA (miRNA) strategy can be utilized to either promote angiogenesis by targeting negative regulators in angiogenesis signaling pathways or inhibit angiogenesis by targeting positive regulators. Nanotechnology and targeted delivery of miRNA or anti-miRNA could be delivered to tumor endothelium using targeted nanoparticles [18]. Lessons learned from antisense technologies and RNA interference approaches will no doubt be relevant in advancement of miRNA therapeutics in addition to the use of nano-targeted delivery of mRNA or anti-mRNA. However, a potential limitation of miRNA-based therapy is the possible off-target effects that might lead to serious adverse effects [19, 20].

Acknowledgement The authors appreciate the generous investment of Richard C. Liebich in much of the work reported in [6, 9–11] included in this chapter.

References

1. Clauss M (2000) Molecular biology of the VEGF and VEGF receptor family. Semin Thromb Hemost 26(5):561–569
2. Alvarez-Garcia V, Gonzalez A, Alonso-Gonzalez C, Martinez-Campa C, Cos S (2012) Regulation of vascular endothelial growth factor by melatonin in human breast cancer cells. J Pineal Res 54(4):373–380
3. Yang J, McNeish B, Butterfield C, Moses MA (2013) Lipocalin 2 is a novel regulator of angiogenesis in human breast cancer. FASEB J 27(1):45–50
4. Chen Y, Fu L, Han Y, Teng Y, Sun J, Xie R, Cao J (2012) Testosterone replacement therapy promotes angiogenesis after acute myocardial infarction by enhancing expression of cytokines HIF-1alpha, SDF-1alpha and VEGF. Eur J Pharmacol 684(1–3):116–124
5. Ribatti D (2010) Erythropoietin and tumor angiogenesis. Stem Cells Div 19(1):1–4
6. Davis PJ, Davis FB, Mousa SA, Luidens MK, Lin HY (2011) Membrane receptor for thyroid hormone: physiologic and pharmacologic implications. Annu Rev Pharmacol Toxicol 51:99–115
7. Mikula-Pietrasik J, Kuczmarska A, Kucinska M, Murias M, Wierzchowski M, Winckiwicz M, Staniszewski R, Breborowicz A, Ksiazek K (2012) Resveratrol and its synthetic derivatives exert opposite effects on mesothelial cell-dependent angiogenesis via modulating secretion of VEGF and IL-8/CXCL8. Angiogenesis 15(3):361–376
8. Kim KH, Bender JR (2009) Membrane-initiated actions of estrogen on the endothelium. Mol Cell Endocrinol 308(1–2):3–8
9. Lin HY, Cody V, Davis FB, Hercbergs AA, Luidens MK, Mousa SA, Davis PJ (2011) Identification and functions of the plasma membrane receptor for thyroid hormone analogues. Discov Med 11(59):337–347
10. Glinskii AB, Glinsky GV, Lin HY, Tang HY, Sun M, Davis FB, Luidens MK, Mousa SA, Hercbergs AH, Davis PJ (2009) Modification of survival pathway gene expression in human breast cancer cells by tetraiodothyroacetic acid (tetrac). Cell Cycle 8(21):3554–3562
11. Lin HY, Sun M, Tang HY, Lin C, Luidens MK, Mousa SA, Incerpi S, Drusano GL, Davis FB, Davis PJ (2009) L-Thyroxine vs. 3, 5, 3′-triiodo-L-thyronine and cell proliferation: activation of mitogen-activated protein kinase and phosphatidylinositol 3-kinase. Am J Physiol Cell Physiol 296(5):C980–C991

12. Gondou A, Toyoda N, Nishikawa M, Yonemoto T, Sakaguchi N, Tokoro T, Inada M (1999) Effect of nicotine on type 2 deiodinase activity in cultured rat glial cells. Endocr J 46(1): 107–112
13. Sudhagar S, Sathya S, Lakshmi BS (2011) Rapid non-genomic signaling by 17β-oestradiol through c-Src involves mTOR-dependent expression of HIF-1α in breast cancer cells. Br J Cancer 105(7):953–960
14. Waugh DJ, Wilson C (2008) The interleukin-8 pathway in cancer. Clin Cancer Res 14(21):6735–6741
15. Volin MV, Koch AE (2011) Interleukin-18: a mediator of inflammation and angiogenesis in rheumatoid arthritis. J Interferon Cytokine Res 31(10):745–751
16. Wu Y, Li YY, Matushima K, Baba T, Mukaida N (2008) CCL3-CCR5 axis regulates intratumoral accumulation of leukocytes and fibroblasts and promotes angiogenesis in murine lung metastasis process. J Immunol 181(9):6384–6393
17. Chauhan VP, Stylianopoulos T, Martin JD, Popovic C, Chen O, Kamoun WS, Bawendi MG, Fukumura D, Jain RK (2012) Normalization of tumour blood vessels improves the delivery of nanomedicines in a size-dependent manner. Nat Nanotechnol 7(6):383–388
18. Murphy EA, Majeti BK, Barnes LA, Makale M, Weis SM, Lutu-Fuga K, Wrasidlo W, Cheresh DA (2008) Nanoparticle-mediated drug delivery to tumor vasculature suppresses metastasis. Proc Natl Acad Sci U S A 105(27):9343–9348
19. Heusschen R, van Gink M, Griffioen AW, Thijssen VL (2010) MicroRNAs in the tumor endothelium: novel controls on the angioregulatory switchboard. Biochim Biophys Acta 1805:87–96
20. Rayner KJ, Suarez Y, Davalos A, Parathath S, Fitzgerald ML, Tamehiro N, Fisher EA, Moore KJ, Fernandez-Hernando C (2010) MiR-33 contributes to the regulation of cholesterol homeostasis. Science 328:1570–1573

Index

S.A. Mousa and P.J. Davis (eds.), *Angiogenesis Modulations in Health and Disease:*
Practical Applications of Pro- and Anti-angiogenesis Targets,
DOI 10.1007/978-94-007-6467-5, © Springer Science+Business Media Dordrecht 2013

Printed by Publishers' Graphics LLC
KSO130517.15.17.18